D1188476

Engaging the Everyday

Engaging the Everyday: Environmental Social Criticism and the Resonance Dilemma

John M. Meyer

The MIT Press
Cambridge, Massachusetts
London, England

© 2015 Massachusetts Institute of Technology

All rights reserved. No part of this book may be reproduced in any form by any electronic or mechanical means (including photocopying, recording, or information storage and retrieval) without permission in writing from the publisher.

MIT Press books may be purchased at special quantity discounts for business or sales promotional use. For information, please email special_sales@mitpress.mit.edu.

This book was set in 10/14 pt Sabon by Toppan Best-set Premedia Limited. Printed and bound in the United States of America.

Library of Congress Cataloging-in-Publication Data is available.

ISBN: 978-0-262-02890-5 (hardcover); 978-0-262-52738-5 (paperback)

10 9 8 7 6 5 4 3 2 1

Published with support from the Rachel Carson Center for Environment and Society.

Rachel Carson Center

ENVIRONMENT AND SOCIETY

Contents

Acknowledgments

One of the real pleasures of finishing this book is the opportunity to give thanks to those whose support, encouragement, and critical engagement have been crucial along the way.

David Schlosberg, Sherilyn MacGregor, and a third anonymous reviewer each read the entire manuscript and provided detailed and invaluable advice. This book is far better as a result. Friends and colleagues in the working group on environmental political theory, which meets annually at the Western Political Science Association conference, constitute a vibrant and growing community that is one of my primary sources of intellectual stimulation and camaraderie. Many participants have read one or more chapters and pushed me to clarify, refine, or rethink my arguments. These include John Barry, Andrew Biro, Sheri Breen, Mark Brown, Peter Cannavò, Bill Chaloupka, Robyn Eckersley, Lisa Ellis, Teena Gabrielson, Cheryl Hall, Breena Holland, Joel Kassiola, Joe Lane, Tim Luke, Michael Nordquist, David Schlosberg, Piers Stephens, Steve Vanderheiden, Justin Williams, and Harlan Wilson.

Foundations for the book were laid during a wonderful year spent as a member of the School of Social Science at the Institute for Advanced Study in Princeton (2004–2005). Thanks to Joan Scott for her encouragement and for bringing together a stimulating seminar on the nature of interdisciplinarity and to Michael Walzer for talks and walks and for urging me to publish some of my reflections on the state of environmentalism in *Dissent*. Among the many others who provided critical insight on my writings and ideas that year, I particularly wish to thank Stephen Gardiner and Victoria Kamsler.

Often relegated to the back burner in the ensuing years, the book came to fruition during another terrific year, this one in Munich as a fellow of the Rachel Carson Center for Environment and Society (2012–2013). Christof Mauch and Helmuth Trischler, as well as their terrific staff, have created a unique and valued space for work in the environmental humanities. Once again, my colleagues provided a great deal of constructive guidance, intellectual stimulation, and fun. With apologies to those I've overlooked, I thank Ellen Arnold, Franz-Joseph Brüggemeier, Lawrence Culver, Rob Emmett, Maurits Ertsen, Eva Jakobsson, Matt Kelly, Tom Lekan, Michelle Mart, Kenichi Matsui, Jan-Henrik Meyer, Giacomo Parinello, Chris Pastore, Maya Peterson, Michel Pimbert, John Sandlos, Hanna Schösler, Nicole Seymour, Don Worster, Frank Zelko, and Tom Zeller.

In 2012, Acadia University in Nova Scotia generously invited me to join them as a Harrison-McCain Visiting Professor. Andrew Biro and his colleagues provided a warm welcome and the students in Andrew's graduate colloquium on social and political thought offered a stimulating set of opportunities to explore the politics of materiality with them.

Two chapters of this book had a previous life as journal articles. A version of chapter 2 was published in May 2011 as "We Have Never Been Liberal: The Environmentalist Turn to Liberalism and the Possibilities for Social Criticism," in *Environmental Politics* 20, no. 3: 356–373. A version of chapter 5 (with a bit that migrated into chapter 4) was published in February 2009 as "The Concept of Private Property and the Limits of the Environmental Imagination," *Political Theory* 37, no. 1: 99–127. Thanks to the reviewers at both journals and to Mary Dietz at *Political Theory* for constructive advice.

In addition to the settings already noted, writing that found its way into the book was presented at the following venues: the Southern Political Science Association conference in 2005, the Environmental Studies Department Seminar at UC-Santa Cruz, the American Political Science Association conferences in 2009 and 2011, the Atlantic Provinces Political Science Association conference in 2012, the Sydney Ideas Roundtable at the University of Sydney in 2012, the Institut für Politische Wissenschaft at Leibniz Universität–Hannover in 2013, the ECPR Environmental Politics and Policy PhD summer school at Keele University in 2013, and the Association for Political Theory conference in 2014.

Despite these forays to far-flung locations, my home for many years has been on the North Coast of California and at Humboldt State

University. I share interests and passions with many of my colleagues and students in both the Department of Politics and the Environment and Community graduate program, making it a richly rewarding place to work. The university also generously supported two sabbatical leaves for this and related projects. To live in a place with both a strong sense of community and such physical beauty is a privilege. There are far too many people here to thank and appreciate, but at the tip of the iceberg are some close colleagues (and good friends!) who have been with me for the long haul and whose work and interests have informed mine: Mark Baker, Yvonne Everett, Judee Mayer, Betsy Watson, and Noah Zerbe.

I couldn't ask for a better publisher than the MIT Press. Professionalism, attention to detail, and commitment to quality characterize everyone I've worked with at the press. First among them has been Clay Morgan. As acquisitions editor, Clay not only has been supportive and encouraging of my work for many years but also has been crucial to the development of the whole field of environmental political theory. I will miss his good judgment and calm influence now that he has retired.

Not all of those I wish to thank fit neatly into one of the previous categories, but their guidance, invitation, or other support along the way have been no less valued: Jutta Joachim, Michael Maniates, Sam Nelson, and Heather Sullivan. Also, thanks for the last minute prodding and help improving the book's title from Bernie Yack, Cheryl Hall, and Mark Brown.

Writing about the practices of everyday life has made me all the more aware of how blessed I am by the love and support of my family. Carolyn's patience, caring, encouragement, and laughter have sustained me through good times and bad; her passion for her own vocation inspires me in mine. Jake and Emelia have grown into thoughtful, caring, talented, and engaged individuals; we are privileged to be their parents. Although work on this book often conflicted with time together, it also created the opportunity for some grand adventures. The book is an outgrowth of a love for them for which my words come up short. Nevertheless . . . thank you.

1

Introduction: The Resonance Dilemma and Environmental Social Criticism

On the one hand, environmental challenges—including climate change—threaten the very fabric of our lives, such that the present course appears literally unsustainable. On the other hand, far-reaching efforts to address these challenges rarely seem to resonate with citizens in the United States and other affluent, postindustrial societies. Let us call this the *resonance dilemma*. President Obama expressed the second aspect of this plainly, arguing rhetorically that "you may be concerned about the temperature of the planet, but it's probably not rising to your number-one concern . . . And if people [i.e., environmentalists] think, well, that's shortsighted, that's what happens when you're struggling to get by."[1] This gap between the broadly felt concern and the lack of priority placed on action is one of the more consistent and long-standing features of US public opinion about environmental challenges and climate change,[2] and it's not just a dilemma in the United States. Summarizing a 2013 survey of public opinion in 22 countries, a spokesperson concluded that although "evidence of environmental damage is stronger than ever . . . our data shows that economic crisis and a lack of political leadership mean that the public are starting to tune out."[3]

Attention to differences of power and privilege allows us to see important dimensions of the resonance dilemma. In the United States, for example, the lens of race is particularly illuminating. Here, nationally prominent environmental and climate leaders remain overwhelmingly white, despite the growth of local environmental justice organizations in poor and minority communities and a rapidly growing nonwhite population in the country as a whole.[4] The explanation for this disparity is not that climate change or other issues have less effect on these groups; the opposite is often the case. The problem is also not that higher

percentages of members of these groups are opposed to measures to mitigate climate change or other challenges; race and class show no consistent correlation with conventional measures of environmental concern
in the United States. Yet many have argued that the lack of diversity in
national leadership contributes to conceptualizing environmental issues
in a manner that fails to connect with the urgent priorities of everyday
life for many in nonwhite and disadvantaged communities.[5] Addressing
this is not easy. It is neither a matter of rising above overt racism nor
simply a matter of more outreach, communication, or "spin." As Ryan
Young, legal counsel of the Greenlining Institute, explains, "If you want
to gain the trust of the emerging nonwhite majority, it's not just a mes-
saging thing. It's a values thing. You must understand the values of these
communities and craft policy around that."[6]

As Young goes on to suggest, an analysis that begins with the values
and everyday concerns of marginalized communities requires reimagining much that is often identified with environmentalism today. Attention
to the emerging nonwhite majority in the United States also helps to
explain recent findings that the most diverse generation—younger adults
of the so-called millennial generation—are significantly less likely than
their elders to identify as environmentalists.[7] Yet an analysis that carefully considers how identities intersect with environmental concern is
relevant within white as well as nonwhite communities—among those
with relative privilege as well as those who are often marginalized.

Consider this: many political theorists and practitioners have argued
for the transformative potential of deliberation to facilitate greater action
on environmental concerns.[8] Empirical evidence exists that this can lead
to changes in participants' understanding and perspective. In the rare
case in which deliberation is structured to lead directly to policy decision
and implementation, this can foster change. Yet where this immediate
link has not been present, the resonance dilemma again emerges. Kersty
Hobson captures this well when describing the outcomes of a deliberative forum on climate change held in Australia. In studying the immediate effect of this multiday event, she concludes that it stimulated significant
"shifts in research participants' perspectives around key issues undergird-
ing climate change."[9] Yet in follow-up interviews conducted six months
later, she finds that these changes were not enduring. The comments of
one initially enthusiastic participant make the reasons plain: "I'm married,

two kids, I went home and gave the most polite and gentlest talk I could to them without freaking them out. My wife was like 'yeah, whatever,' and two days later it [the subject of climate change] left the table, and that's happening to me, and I'm committed. So what do you do?"[10]

Observations such as these could be multiplied, but all point toward the same dilemma: the way in which environmental problems—and environmentalism itself—are currently understood does generate expressions of concern but does not resonate deeply with many portions of the population. As a result, it cannot generate the political context that is a precondition for decisive democratic action and change.[11]

Identifying the resonance dilemma as the key impediment to effective action can be contrasted with another familiar diagnosis, which focuses instead on the role of virulent opposition.[12] Especially with regard to climate change, this takes the form of "denialism." To be sure, denial of the reality of human-induced climate change is promoted by a powerful and interlocking network of industries, conservative foundations, and right-wing think tanks.[13] Denial of climate science and action has fed off of—and contributed to—increasing polarization in US politics and has often poisoned scientific and political discourse. By contrast, its influence in most other countries has been more muted.[14] Yet even in the United States a singular focus upon opposition and climate science denialism as the obstacle to change is itself a form of denial; it imagines that these efforts have been far more monolithic and effective in shaping public opinion than is the case.[15] In this way, it suggests that overcoming denial—increasing the numbers of those who accept the reality of anthropogenic climate change—would lead to decisive public action. In fact, a consistent majority in the United States already *does* accept the reality of climate change as a problem,[16] but the lack of resonance remains a potent impediment here, a dilemma even more apparent in countries in which active opposition is less conspicuous.

Certainly there are those—likely including many readers of this book—for whom climate change, biodiversity loss, toxic pollution, and other challenges to sustainability are urgent priorities at the forefront of daily decisions. Let's call these members of the population active "environmentalists." The starting point for this book is the conviction that these challenges are too big and therefore unlikely to be addressed effectively if left just to these environmentalists. Moreover, environmentalists are unlikely to cultivate the broader and more deeply resonant concern needed by simply adjusting tactics or shouting louder.

What Is to Be Done?

If we take the resonance dilemma seriously, what should be done? One familiar response is: *wait*. President Obama's comment, noted at the outset, is one manifestation of this rather fatalistic response; for people "struggling to get by," he asserts, these challenges are unlikely to be a priority. The suggestion that we must address this struggle first, waiting for a more opportune moment to address climate change powerfully, seems to be implicit. In making this comment, the president referred to the lingering damage and tepid recovery from the so-called Great Recession of 2008. Yet in fact, environmental challenges have rarely been *prioritized* by a broad public. For example, in polling data of Americans from 2001 to 2013—encompassing both economic booms and busts— the relative priority ranking of "protecting the environment" and "dealing with global warming" changed very little.[17] If we broaden our sense of the reasons that people "struggle" to include preoccupations with time-consuming matters connected with children, family, friends, work, school, home, and other aspects of the lives that consume most of us every day (the sorts of preoccupations reflected in survey results' top ranked priorities for action), then it will be quite rare and fleeting for environmental challenges to trump everyday concerns and rise to the top of many people's priority list.

It makes little sense, then, to wait for full employment, a more bullish market, or a more settled and secure society with the expectation that environmental concerns will then emerge as a top priority. More broadly, we should not expect that most people either will or should prioritize environmental action if that is understood to require them to overcome or transcend their preoccupation with their families, livelihoods, homes, and other aspects of everyday life. Rather than *trumping* everyday preoccupations, we would do far better to explore ways to address climate and environmental challenges as *an integral part of* these concerns.

This alternative project requires that we confront limitations inherent in dominant framings of the environmentalist agenda. This agenda is often advanced with confident appeals to the authority of science and morality, drawing upon one or both for an urgent directive to act. To be sure, both science and morality must deeply inform the quest for sustainability. Yet no matter how sound the science nor how widely shared the moral propositions, they cannot resolve differences in how to act or in the distribution of consequences from a given act.[18] The urgent demand to act cannot erase—and should not occlude—the inescapably political judgments that shape what action is deemed fitting, feasible, or fair in a given context. To confuse scientific expertise or moral conviction with this political judgment can often lead to perceiving resistance or inaction as objectively ignorant, apathetic, or immoral egoism.[19] This perception, in turn, contributes to the view held by many that environmentalists are arrogant or elitist.[20]

The alternative is to imagine an agenda for environmental sustainability that emerges from everyday concerns and is thus more deeply resonant with the lives of those of us who—in one way or another, and not just economically—"struggle to get by." By taking the practices of everyday life seriously we are also better positioned to avoid the elitist or arrogant tone that can obstruct the resonance of environmental criticism. This book is an effort to think through the contours of this alternative— to theorize both some of the obstacles that stand in its way and some of the opportunities that it might open up. I do so by engaging practices and values that resonate widely in affluent, postindustrial societies, with the aim of fostering a more expansive political imagination.

The practices that I focus upon are related to use of land, dependence upon automobiles, and dwelling in homes. Each is a "material practice" in the sense that human experience is inextricably interwoven with technology, the built environment, and the nonhuman world. Each is also often constituted as a private realm, yet these practices shape public ideas and values in profound ways. Engaging them thoughtfully requires openness to the ways in which these practices both enable and constrain present lives and livelihoods. This must become central to the task of environmental social criticism.

Neither Outside *Critics* nor *Inside* Players . . . *Inside Critics*

In this book, I regard visions of, and arguments for, environmental sustainability as social criticism. That is, they are offered as criticisms of existing societies with the intent—directly or indirectly—of promoting change. Some are convinced that to escape the assumptions and biases of the society the critic wishes to see changed she or he must be emotionally detached from that society, approaching it from the perspective of an uncompromised—and uncompromising—outsider. From this position, the critic speaks as a prophet from on high, condemning the existing direction of society and the self-absorption or preoccupations of its citizens. This critic appeals to the sense of righteousness among the minority who share her or his view, while at the same time calling upon the broader citizenry to wake up, confront their illusions, and change their lives, offering a new vision for the social order.[21] Many radical environmental thinkers fit this model of criticism, presenting themselves as detached from and seemingly outside of the culture they seek to change. Deep-green resistance leader Derrick Jensen plays the role especially well. Among what he terms the "premises" of his two-volume work, *Endgames*, are that "the culture as a whole and most of its members are insane" and "it is a mistake (or more likely, denial) to base our decisions on whether actions . . . will or won't frighten fence-sitters, or the mass of Americans."[22]

Outside critics like this frequently attribute some sort of "false consciousness" to many members of the society, a consciousness that only the critic and his or her supporters are able to recognize as false.[23] It should not be surprising that in many cases the critic's claims are thus rebuked as patronizing or paternalistic by members of society; after all, the very basis for this criticism is a dichotomy between those who know and the many who don't or those who see clearly and the many who are myopic. The critic's detachment from others in society makes it easier to condemn the attitudes and behavior of the masses and reject what the latter regard as their ordinary good sense and priorities. From the perspective of the populace at large, in turn, this critic's evident disdain invites their rejection.[24]

It ought not be surprising, then, that with regard to environmental challenges this critic often relies upon the notion that large-scale ecological collapse or catastrophe will be necessary (and likely) to break the spell, wake people up, and change the direction of society; only when faced with "the end of the world as we know it" will the relative inconsequence of our everyday preoccupations become apparent.[25] The perceived rigidity of contemporary social norms means that anything less is unlikely to provoke the necessary change. Some outside critics might take refuge in the conviction that they are right and others are wrong, but given that criticism aims to create change the inability to connect with a broader audience can only be regarded as failure.

Often, at least in nominally liberal-democratic societies, the only alternative to the model of the social critic as moralistic outsider is believed to be the issue advocate working inside of existing corridors of power to promote policy or regulatory change through persuasion, deal making, and compromise among decision makers. The latter is the instrumental role envisioned by many within large NGOs and governmental agencies devoted to environmental concerns. It eschews the finger-wagging moralism and paternalism of the outsider in favor of the quest to win over some of Jensen's "fence-sitters." This will have to be accomplished, therefore, without challenging influential social norms or threatening dominant institutions. Of course, this approach has also long been subject to criticisms, including charges that such players are rarely willing to rock the boat (which might jeopardize their own insider status) or to acknowledge the extent of the change necessary to address the problem effectively. Nonetheless, the contrast here is clear. The alternative to being an outsider, a critic tilting at windmills, is to be an insider, a player getting his or her hands dirty while pursuing changes within the system.

Central to my analysis in this book is the conviction that the contrasting ideal types of the "outside critic" and the "inside player" do not exhaust the range of viable alternatives. The social critic need not position him or herself as an outsider. More to the point, the likely success of critics depends crucially upon their ability to speak in a manner that resonates with citizens while simultaneously arguing for extensive, meaningful change from the status quo. Those who position themselves as outsiders will be hard-pressed to connect their criticism with citizen concerns and are especially poorly situated to do so. Conversely, inside players are poorly situated to question the rules of the game they are playing. A third model, that of the immanent or engaged social critic, draws on elements of both the outside critic and the inside player. I will term this third model "inside criticism" to highlight the elements it draws from each of the others. Consistent with the former, inside criticism offers critical distance from the status quo and a reinterpretation or reimagining of existing practices and ideas. Consistent with the latter, it offers a pragmatic engagement with present realities. It can do both, I argue, only by introducing something new: a critical focus upon familiar everyday material practices.

That, at least, is the intention. As vital as I argue that insider criticism is, it is surely not easy to realize. Integral to this model is the notion that critical distance can be achieved without positioning oneself as an outsider. To do so, society must be conceived as potentially more malleable and less rigid than is often imagined. Michael Walzer, a key proponent for this sort of model of engaged social criticism, captures one element of the skeptical response well. Skeptics retort: "Don't the conditions of collective life—immediacy, closeness, emotional attachment, parochial vision—militate against a critical self-understanding? When someone says 'our country,' emphasizing the possessive pronoun, isn't he likely to go on to say 'right or wrong'?"[26]

In this imagined instance, the critical role of the insider collapses. Although I don't wish to minimize the enormous challenges it faces, I argue with Walzer that critical self-understanding is possible and that this type of "inside criticism" is most in keeping with the aspirations for democratic change. To take the possibility of this type seriously, however, we must recognize not only its distance from the "outside critic" but also its distance from the "inside player." That is, an inside critic begins by engaging the everyday practices and experiences of members of her or his society but is distinguished by the willingness to *criticize* these in terms that can be heard by these same members.

A case for this connected, engaged, or immanent form of criticism has been made by a variety of theorists. Among political theorists, in addition to Walzer, James Tully's recent work on public philosophy is valuable here; on a longer horizon, Antonio Gramsci's call for organic intellectuals is also notable.[27] My aim, however, is not to develop a critical genealogy of this type of criticism. The proof of the pudding must be in the eating. The case here will be provisionally developed through theoretical critique and analysis, but then will be manifest through reflection upon actual material practices in the second part of this book.

A question that remains in this contrast is: inside or outside *of what*? It is common to assume that the "what" is an ideology, worldview, or ontology that represents a frame or horizon widely shared by members of a given society. Often "liberalism" or "modernity" are regarded as the frame either that one embraces and acts from within or that one must escape and act or speak from without. Yet Richard Flacks persuasively argues:

Rather than ideologies, Americans are overwhelmingly committed to their lives—by which I mean, of course, not simply survival as such, but the roles, relationships, and purposes that constitute daily round and personal identity. That commitment is the bedrock of popular ideology in American society—the framework people are most likely to use in reacting to political leaders, events, and issues. Apparent inconsistencies in individual belief and attitude tend to become more "logical" once their connections to the maintenance and development of the individual's everyday life are grasped.[28]

I am convinced that this point applies beyond just Americans. Thus the task of inside social criticism does not require it to be positioned within an ideological framework defined as liberalism or modernity but instead to recognize and engage the "roles, relationships, and purposes" central to daily life. To focus on everyday life in this way is to go beyond an isolated realm of ideas and values to reflect upon the material practices that constitute our lived experiences. In the next chapter, I develop this point and argue that the treatment of liberalism in this manner by many contemporary environmental thinkers—illuminating though it can be in many particulars—is a category mistake when the aim is to critique unsustainable aspects of contemporary society and lives.

Appreciating the three-way relationship between outside criticism, inside players, and inside critics also allows for a fresh take on the familiar distinction between radicalism and reform. Radical thinkers are often styled as those who transcend or reject extant values and experiences of everyday life in the mold of outside critics. They call for "total revolution" or proclaim the necessity of embracing a completely new worldview.[29] Tactically, radicalism is often identified with appeals to civil disobedience, disruption, and in some cases (but by no means most) violence. It is often regarded as inconsistent with efforts to promote legislative change or other tactics that seem to require acceptance of compromise.[30] From this point of view, insiders can never be more than piecemeal reformers, accomplishing little more than rearranging deck chairs on the Titanic. In this sense, the radical-reformer dichotomy tracks closely with that between outside critics and inside players.

Yet radical thinking—literally getting to the *roots*—can and should be understood in a very different sense. If we conceive the root problem or predicament as I have at the outset—as the contemporary lack of resonance of grave environmental challenges among many citizens—then the radical must seek out and address the roots of this resonance dilemma, something that can only be addressed by the critic as *insider*. Such radical thinking might be consistent with a wider variety of strategies and tactics for action; the criteria for choice are dependent upon pragmatic judgments about the act's likely success in achieving the radical goal of greater resonance with the everyday concerns of the public.[31]

In conclusion, resonant social criticism is a valuable and, I maintain, *necessary* condition for meaningful democratic change. Given the vast inequalities of power, privilege, and influence in existing democratic societies, however, such criticism is not a *sufficient* condition. Many have drawn attention to consequences of these inequalities; others have argued for alternative institutions and processes of representation and deliberation that might better facilitate environmental action.[32] My focus in this book ought not to be read as diminishing the importance of such arguments and approaches, yet I do wish to emphasize that the need for thinking carefully about social criticism is not diminished by these.

Environmental Political Theory

Here I want to note some characteristics important to my inquiry that are widely shared by environmental political theorists, as well as some other aspects of my approach that are more distinctive. The label "environmental political theory" (EPT) is useful as a signifier of a vibrant and growing field of inquiry with a shared set of substantive concerns and an extended family of conventions, methodological approaches, and bodies of literature that inform our work. It is not, however, a shared ideological framework nor a shared conviction about strategies for social change.[33] Moreover, certainly not all EPT is envisioned by its authors as either a form of or an account of social criticism in the manner I've described it here.

Environmental political theory can also be understood in less aspirational terms—as subfield of a subfield of an academic discipline. Here, "environmental" is an adjective modifying "political theory," where the latter is regarded as a specialized (and oft-marginalized) subfield of the discipline of political science. The contours of environmental political theory have been crafted over the past two decades in a time of hyper-professionalization, with increasing academic specialization moving full-speed ahead. Certainly, there are ways in which EPT is a reflection of this specialization and professionalization. Yet in other ways its very constitution resists it. It is the latter that most interests me here.

Whereas political theory sets the human political community as its purview and a critical and normative analysis as its aim, EPT broadens that already wide horizon to the entire nonhuman world within which this political community is embedded while retaining the critical and normative commitment. Political theory is an undisciplined subfield of the political science discipline, which regularly consorts with other fields, including intellectual history, philosophy, social theory, gender studies, and cultural studies. At times, political theorists also draw upon empirical accounts of public opinion, communication, and power relationships from various disciplines. EPT follows these leads and also integrates insights from ecological science, geography, environmental history, science and technology studies, and other fields of inquiry that allow us to understand better its more expansive horizon. Environmental political theory is thus inescapably, and broadly, a cross-disciplinary pursuit.

The result is work that is often distinguished by its eclecticism and its amateurism. This might seem like an all-too-frank admission that will allow scholars in other fields to deride and dismiss the work that results, yet I wish to challenge that appearance. Given the pressures toward professionalization and narrow specialization of academic work, the opportunity to pursue work that aspires, in part, to escape constraining disciplinary boundaries and the narrow strictures of scientism is valuable and necessary.

To explore wide swaths of the human experience, one is necessarily an amateur and eclectic. The alternative is not to do so as a professional, but rather not to do so at all. One can nonetheless apply standards of care and reflection to a wide-ranging inquiry. That is what I believe distinguishes much work of environmental political theory. To be clear, it would be disingenuous for me or others in my field to fully claim the mantle of an amateur. I have a PhD, a tenured job in the American academy, and my writing reflects the style, insights, and blinders of my profession—even when it does so by omission. In that sense, I am a creature of my subfield of a subfield of a discipline. Amateurism and eclecticism, in this context, is a modest but self-conscious form of resistance against this professionalization and specialization; political theory in general, and environmental political theory in particular, is just amorphous enough to allow for this meaningful form of intellectual freedom.

Michael Walzer has characterized this freedom of political theorists to resist professionalized and specialized norms dominant in the academy as a "political theory license."[34] Although skeptics might characterize such a license as permission to dabble and so to create work that is not "state of the art," I think the best response to such a charge is to recognize that trade-offs between depth and breadth, between sophistication and accessibility, and between the fine details and the big picture are inherent in all forms of inquiry. My aim, exhibited in the best work of political theory in this register, is to achieve a greater balance between these extremes in an era in which the pressures upon academic work tend toward the former end of this spectrum.

To the extent that there is a shared agreement among environmental political theorists, it is that environmental challenges cannot be understood by scientists alone and that no nonpolitical means of addressing them is possible. This is not a wholesale rejection of technology or of economic incentives but a recognition that these are also dependent upon judgments, choices, and collective decisions that are manifestations of politics broadly construed. The alternative to politics is neither technology nor markets, but warfare or magic—which is to say, there really is no alternative. Similarly, there is a shared sense that political theory, with its attention to macrolevel phenomena, is a necessary corrective to approaches that focus primarily or exclusively upon individual-level ethics and choices.[35]

I am inspired by the work of my colleagues in this field, have learned a tremendous amount from it, and share the commitments I've described here. Yet in this book I take leave from other approaches also often found in this body of work. Unlike many, I do not reinterpret liberal or modern political philosophy—or any other philosophical tradition—nor do I begin with an analysis of climate change, biodiversity loss, or other specific environmental issues. Instead, I draw upon a variety of theorists for the insight needed to delineate a framework for examining material practices. I then use this framework to discuss practices that are important to contemporary conceptions of property, individual freedom, and citizenship, and thus central to contemporary environmental challenges.

Critical and normative theory can and should aspire to illuminate characteristics of the society that is the subject of theoretical inquiry. Yet too often we act as though rereading and engaging a given tradition of theorizing is a means of engaging with the society itself. Living in "modern" society, for example, many scholars treat theories of modernity as an apt and comprehensive characterization of actual society and so regard critical analysis of such theories as akin to a critical analysis of society itself.[36] This sort of approach treats theory unproblematically as the distillation of practice and so as a surrogate that can serve as an appropriate object of critical engagement.

This treatment leads to the misguided expectation that critiquing an influential theoretical tradition is concomitant with effective social and political criticism. Treating a theory in this way reifies rather than examines its relationship to popular opinions and widely shared cultural constructions. Problems in society thus come to be construed as a product of the theory itself. The resulting project requires that we either reject the theory, replacing it with another (somehow transforming a broadly shared worldview or ontology), or reinterpret the theory in a more expansive or elastic manner. This approach entails a critical engagement with theory, presuming that a reworking of a theoretical tradition is itself a form of critical engagement with society. By making this presumption explicit, we can better understand the motivation for work in this vein. It also draws attention to the presumption's fragility and so to aspects of this approach that are far from self-evident—either unconvincing or in need of an explicit defense.

In the next chapter, I focus on the treatment of liberal theory among many environmental thinkers—including political theorists—to argue that this treatment confuses its relationship to practice. The same treatment can be found among those who reject the liberal tradition and the growing number of recent theorists writing from within it. Rather than take sides (i.e., for or against liberalism or for or against a particular interpretation of liberal theory), I argue that this treatment itself is a category mistake. It both overestimates the role of theory in the constitution of present practices and misconstrues the role that theory might play in facilitating a rethinking and restructuring of these practices. My critique here is shaped by insights from scholars who include Karl Polanyi and Bruno Latour, both of whom have drawn our attention to the role that theory can play in obscuring the complexity of actual practice.

I aim to advance the case for an alternative, pragmatic role for theory— not as frame or shaper of the world but as reflection upon and interpretation of everyday experience in all its complexity and contradictory nature. John Dewey identifies well the appropriate criteria for effective theorizing and its relation to practice:

Does it end in conclusions which, when they are referred back to ordinary life-experiences and their predicaments, render them more significant, more luminous to us, and make our dealings with them more fruitful? Or does it terminate in rendering the things of ordinary experience more opaque than they were before, and in depriving them of having in "reality" even the significance they had previously seemed to have? Does it yield the enrichment and increase of power of ordinary things which the results of physical science afford when applied in every-day affairs? Or does it become a mystery that these ordinary things should be what they are; and are philosophic concepts left to dwell in separation in some technical realm of their own?[37]

I utilize this pragmatic role for theory later in the book to reflect upon everyday dilemmas and practices. Certainly, past theoretical traditions have done much to shape these practices; nothing I write here should suggest otherwise. But noting that there might be other elements of past theory that are critical of these practices does little to help us imagine alternatives to them, or to develop resonant critiques of them, except to the extent that these are grounded in the practices themselves. In sum, my aim here is not to make an antitheoretical move but to advance a conception of theory that illuminates practice, because it is directly attentive to the practices under consideration.

Everyday Practices and Political Concepts

In examining everyday practices, I do not formulate policy advice or specific strategies for social change. Instead, I seek to develop an understanding of these practices that is sufficient to offer perspective and insight into our political concepts and values, thus providing a grounded basis for reshaping our talk about these subjects. I consider three practices (or, more precisely, areas of practice): land use, transportation by automobile, and home dwelling. Described as a *practice*, each of these goes beyond the limitation of arguments that characterize change as merely the aggregate product of autonomous individual choice. By the same token, attention to practices also avoids reifying structure as the determinant of individual behavior. Instead, attending to practices draws our attention to the context that structures individual choices and the collective possibilities for such choices to modify this context.[38]

Certainly, these three areas do not offer comprehensive coverage, nor are they indisputably the most important; there are other practices that could also have been chosen. Food, for example, is enmeshed with a great many practices with enormous environmental consequences, a reality that may finally be getting the recognition in both the academy and popular media that it deserves. My claim—a relatively uncontroversial one—is only that the practices of postindustrial society that I attend to here do have an immense impact upon human and nonhuman environments.

In each of the areas of practice that I consider, there are consequences for our political concepts and values. I focus on one political concept in relation to each practice. Controversies surrounding the use of land, for instance, invoke an array of vital political concepts, including freedom, justice, and equality. Central to all, however, is property—especially *private* property—itself. Thus it is the concept of property that becomes the focus of my reflections upon land use. Although many begin with a supposedly liberal theory of property and then seek to stretch or challenge that concept in order to foster more sustainable land-use practices, I argue that neither land-use practices nor law conform to this theory of property now nor have they in the past. Recognizing the good reasons for this, we can begin to reimagine and reconceptualize what property might mean and ought to mean.

Similarly, the heavy reliance throughout much of the world today on the privately owned automobile as a primary means of mobility has substantial implications for the ways in which we think about freedom. Although some might view this widespread practice through the competing lenses of liberal and republican conceptions of freedom, or negative and positive conceptions, I begin instead by reflecting upon how freedom is understood to be advanced by a society rooted in automobility as a basis for then asking where and how this notion of freedom might also be understood to be constrained by the practice of automobility.

Finally, home dwelling—both the building of houses and the practices of households—is clearly central to contemporary patterns of consumption. Of course, where and how these houses get built also has much to do with our concept of property; living in a home has much to do with our concept of privacy and freedom. Yet in this chapter, I focus upon home dwelling through a familiar contrast between a sphere of consumption and a sphere of citizenship. I conclude that a concept of citizenship that excludes experiences in private homes and households from its purview is an emaciated one that fails to recognize everyday practices with significant promise for environmental sustainability. However, one that simply equates being an environmental citizen with the choice of products we consume undermines the distinctive power and value of citizenship.

My engagement with areas of everyday practice is not strictly instrumental—that is, it is not "merely" a means to address the drivers of climate change and environmental impact. The inadequacy of our theoretical concepts to capture our actual material practices offers an intrinsic reason to address this gap, because there are elements of these practices that might be drawn upon as a basis for criticizing the status quo and advancing resonant arguments for social change.

All too often, public arguments for change are countered by the claim that people in postindustrial societies are unwilling to sacrifice anything to address environmental challenges.[39] As one scholar pithily puts this, we should not "expect a sudden moral epiphany that clashes brutally with contemporary lifestyles . . . it is unlikely that citizens abandon their smartphones in order to embrace the charms of a more embedded rural lifestyle."[40] Another asserts that "not only do the already rich people in industrialized countries value their current lifestyle . . . [but] billions in the developing world also aspire for a decent material living standard . . . these global aspirations to modernity make it impossible to win popular support for any radical green programme."[41]

There's a significant insight in these claims that should not be casually dismissed. Yet the premise that seems to underlie this sort of retort is that people in postindustrial societies live in something like the best of all possible worlds now, so there is little motivation for most people in these societies to alter "contemporary lifestyles."[42] Similarly, these authors assume that for those outside these societies to achieve a "decent material living standard" requires mimicking the particular development trajectory of the global north. These assumptions are overstated and overgeneralized.

Reflecting upon everyday practices can help us to identify those spaces where it is least persuasive and arguments for change the most likely to resonate.[43] For example, I argue that reimagining automobility in contemporary society must begin by recognizing the distinctive ways such a society enables a sense of individual freedom. Otherwise, ideas for reducing cars and driving will rightly be resisted as a paternalistic threat to this freedom. Yet the relationship between automobility and freedom is not unidirectional. Attention to freedom also enables greater clarity about the many ways in which dependence upon cars also constrains freedom. In this way, we can reorient our analysis toward changes to our practices and the built environment that would reduce this dependence. Here, the primary goal might be to enable greater freedom *from* automobility. For drivers, this could mean freedom from expensive car and insurance payments, from being stuck in traffic, or from needing to drive for daily provisioning and childcare responsibilities. For those who can't drive or don't own a car, freedom from automobility can have a somewhat different meaning, including greater freedom from reliance on others, from unsafe streets, or from social isolation. A more complex and practice-based account of the relationship between cars and freedom does not determine answers, but it does open up possibilities that are otherwise all-too-readily dismissed as infeasible.

Engaging in material practices is something we all do every day. Reflecting upon them, in the pursuit of sustainability, is something theorists can and ought to do far more often. It would be a mistake, however, to imagine that by doing so we might fix upon a singular, determinate end. It is precisely when our material practices lead to *differences* that *public* issues—and environmental *politics*—emerge. This is despite the fact that material practices are often characterized as *private*. The public and political is not a space contrasted with the private but is a quality of those everyday acts that affect us collectively and help to constitute our world. Although I simply assert these claims here, they emerge from the analysis that I develop in the chapters of part I of this book.

What about the Rest of the World?

The motivation for this book is shaped by global concerns, but the book itself is quite evidently not global in scope. My theoretical analysis is primarily informed by the western tradition of political theory and, more particularly, recent Anglo-American literature on environmental politics and ideas. My discussions of everyday practices are particularly informed by cases, examples, and my own lived experience in the United States, though I also draw upon analyses and examples from the United Kingdom, Canada, Australia, continental Europe, and other postindustrial societies. I do not draw explicitly upon non-western cultural traditions here, nor do I draw meaningfully upon traditional or indigenous forms of knowledge. Although these limitations might be perceived as either false universalism or parochialism, they are an outgrowth of the understanding of social criticism and theory that I seek to advance.

The ecological and social consequences are global, but the impact— especially on a per capita basis—remains disproportionately a result of the material practices of the modest fraction of the global population that lives primarily in affluent, postindustrial societies.[44] Both the level of material consumption and the social context within which this consumption occurs are globally distinctive. I concur with Andrew Dobson that this huge ecological footprint carries with it an equally large and asymmetrical obligation to change our practices,[45] yet I am skeptical of the political efficacy of proclaiming this a citizenly obligation, as Dobson does, and seek to approach these same people and concerns from a different angle.

Attending to the material practices of everyday life in postindustrial societies can also provide a vital corrective to the conceptualization of environmental sustainability in these societies. As I discuss in detail in chapter 3, environmental movements are most commonly viewed as the product of a values shift in so-called postmaterialist societies. By contrast, movements for environmental justice—first emergent in the United States but now increasingly used as a frame for understanding resistance movements that Ramachandra Guha and Juan Martinez-Alier have described as the "environmentalism of the poor"—have made much deeper and more explicit connections to the material flows of water, energy, forests, and other resources that are integral to their livelihood and communal sustainability and resilience.[46] Thus a focus on the material practices prevalent in postindustrial societies ought not to be seen as a manifestation of privilege or exclusivity but as a corrective in those parts of the world that are both the greatest contributors to ecological impact and the areas in which efforts to foster change are least commonly conceptualized in this manner.[47]

Finally, my focus upon practices in postindustrial societies reflects the model of social criticism that I advocate here. Engaged criticism cannot take a view from nowhere or the view of the outsider. I explore and engage with practices that are close to home—both literally and metaphorically—and I draw upon and seek to develop lines of criticism that emerge from these milieus and that are likely to find a home there. Perhaps this sort of argument for change from within will prove insufficient, but I believe it is unquestionably necessary.

Outline of Subsequent Chapters

This book is divided into two parts. In part I, I develop my theoretical approach over the course of three chapters. In part II, I devote three additional chapters to the exploration of material practices in the manner suggested by the argument in the first part. The two are integrally related, and the broader conceptual agenda for a resonant environmentalism relies heavily upon the argument in the first part. Nonetheless, for readers impatient to appraise the "payoff" of this approach or for those less interested in their political theoretical justification the chapters in part II can be read independently.

Chapter 2, "We Have Never Been Liberal," argues for taking leave of arguments about liberalism that have preoccupied many environmental thinkers. My aim is not to reject liberalism per se but to reject the accuracy and efficacy of approaching contemporary society as though it were the encapsulation of liberalism or any other singular political ideology. This leave-taking is a necessary precursor to the exploration of everyday practices that I pursue. My chapter title gestures to Bruno Latour's book *We Have Never Been Modern*, because, like him, I am convinced that viewing society through an encompassing lens such as theories of modernity or liberalism blinds us to much that is particularly worth seeing.

In chapter 3, "The Question of Materiality in Environmental Politics," I navigate the complex relationship between environmentalist movements and ideas on the one hand and material conditions on the other. Political scientist Ronald Inglehart's so-called postmaterialist values thesis captures—in often precise and revealing language—a deeply flawed way of understanding the basis for environmental concern as values oriented and nonmaterialist. This thesis is echoed in a wide variety of contemporary environmental claims, yet even Inglehart has acknowledged its weakness in this context. The so-called new materialism that has recently emerged among some political theorists, and in a plethora of other academic disciplines, is promising, because it seeks to upend the divide between an objective realm of determined and lifeless materialism and a subjective realm of vitality and values. Yet I show that this field must be traversed with care if we hope to gain a constructive way of engaging everyday practices and avoid being drawn into an ontological vortex that returns us to the duality we seek to escape.

Having now set the stage for engaging material practices, chapter 4, "Private Practices yet Political," grapples with the familiar characterization of material practices as private ones. A conceptual dichotomy between the private on the one hand and the public and political on the other remains widespread—not only in liberal philosophy—and is the final conceptual puzzle to be navigated in this section. I do so by drawing insights from a variety of approaches, including feminist theory and a strain of pragmatism that I find in the work of John Dewey and Karl Polanyi.

In part II, I turn to the three cases sketched earlier. Chapter 5, "Land and the Concept of Private Property," reflects upon contemporary and historical controversies over land use and land-use law to challenge and rethink an absolutist concept of private property ownership. Chapter 6, "Automobility and Freedom," first sketches a holistic approach to thinking about autos, drivers, roads, and the industries and cultures that are integral to them—automobility—as a basis for recognizing its inescapable centrality to contemporary societies, including our political values. Then I develop and critique influential notions of what I term "autofreedom." In chapter 7, "Homes, Household Practices, and the Domain(s) of Citizenship," I challenge a familiar account of "green" home and household practices situated in the realm of consumer choice to argue for the potential of such practices—ones central to the reproduction of everyday life—to expand our notion of citizenship.

A commonality across my reflections upon each of these cases is the quest to use them to press outward at received political conceptions—stretching our political imagination and so expanding our sense of the practices that might resonate with people in contemporary, postindustrial societies. Given the plurality of both practices and political conceptions, the arguments that I advance should be read as an invitation to further reflection, discussion, and debate on these themes. At the same time, I seek to ensure that our attention is drawn to the material practices within which we are enmeshed. In this way, it is nothing more or less than an invitation to the *politics* of environmental sustainability.

I

Toward a Political Theory of Materiality

2

We Have Never Been Liberal

There has been a striking turn toward liberalism—understood as an expansive philosophical tradition[1]—among Anglo-American political theorists in recent years. Whether the focus is inequality, community, feminism, multiculturalism, critical theory, or postmodernism, Ruth Abbey notes convincingly that "conversations between liberals and their critics have been superseded largely, albeit not wholly, by arguments among those who place themselves under the liberal umbrella."[2] In this chapter, I detail a comparable turn among many environmental political theorists and assess its significance. This turn gained steam a bit later than within other varieties of political theory but is now well established.

My argument, however, is neither for nor against the possibility of "greening" liberalism per se. I argue instead that the preoccupation with liberalism in this context—being for *or* against—is a category mistake. It is based upon the reification of liberalism as not just a political philosophy but an apt characterization of citizen values and practices in contemporary liberal democratic societies. I argue that neither liberalism nor any other "ism" can offer this; it is in this sense that I argue we have never been liberal. As a result, we would do better to consider directly when and how environmental concerns might resonate with citizens not whether or how they can be reconciled with a particular conception of "liberalism." In short, I conclude that many of the goals that theorists have pursued through their fixation on liberalism are better pursued by taking leave of it.

Pre-dating this turn, as Abbey's account suggests, it was more commonplace for environmental theorists to be severe (outside) critics of liberalism. In this chapter, I consider both environmental theories that reject liberalism and others that embrace it. Superficially, these appear to be polar opposites. Yet in important respects they are two sides of the same coin; their shared preoccupation with liberalism is a commonality that merits our attention. Motivating this examination, then, is the question of *why* these theorists are so focused on liberalism itself. To answer this, I pose two further questions: First, what *work* is the engagement with liberalism meant to *do?* Second, and more broadly, what do these theorists understand liberalism *to be?*

Consider the turn to liberalism among environmental theorists. What work is this turn meant to do? Three possible responses are worth considering. Although they are analytically distinct, an author may be motivated by more than one. First, the turn can reflect a prior normative commitment to liberalism combined with a desire to address ecological concerns. We might imagine the theorist's perspective as: "I'm a committed liberal who is concerned about environmental devastation. My aim is to reconcile my commitment with my environmental concern." Marcel Wissenburg's *Green Liberalism* is exemplary here; in it he argues that liberal theorists "need to address both environmental issues themselves and the theoretical framework of those who 'frame' environmental issues."[3] If one begins with this liberal commitment, then the central question is *how* to be a "green liberal," a question that is the explicit focus of other works.[4] There really is no environmentalist turn *to* liberalism here; it is more accurate to describe this as a turn *by* liberal theorists *to* environmental concerns. Although this allows us to make sense of some instances of the liberal-environmental engagement, it cannot make sense of why there has been a "turn" or shift toward liberalism among those writing environmental theory. Moreover, to the extent that these theorists bring their commitment to liberalism with them—rather than argue for the unique desirability or distinctive ability of liberalism to promote environmental sustainability—they will be persuasive only to those who share these commitments.[5]

Second, the turn can reflect a desire to engage the mainstream of academic political theory, which is at least perceived to be liberal. Richard Dagger illustrates this when he argues that "liberalism in one form or another now seems to be the dominant position among political theorists, and a book of this sort [on political theory and the ecological challenge] will have to speak to liberal concerns if it is to have any practical effect."[6]

Notice here that in characterizing the "dominant position" of liberalism Dagger refers to its status "among political theorists." In this sense, the argument is a quest for recognition by—and dialogue with—one's academic peers. Yet this motivation simply invites the same question to be posed about other political theorists: what accounts for *their* turn to liberalism, as Ruth Abbey has described it? Academic conventions and professional pressures might offer some insight here. Yet if treated as a singular explanation, this would cast political theory as little more than a slave to fashion. That would be overly cynical. The sense that "everybody's doing it" may be a factor, but in itself it is an insufficient explanation. Indeed, Dagger's claim also points beyond contemporary academic fashion, arguing that environmentalists must speak to liberal concerns "if it is to have any practical effect." Concern with such effect is central to a third answer to the question of what the liberal turn is meant to do.

Finally, the turn can reflect a desire to theorize ways that environmental arguments are likely to resonate with a broad public in liberal democratic societies—and a related conviction that nonliberal forms of discourse will fail to do this. In this sense, theory is conceived as a direct or indirect form of engagement with practice, aimed at understanding the conditions for—and in some cases promoting—change within these societies. From this perspective, political engagement is likely to be effective only if it takes place within a liberal idiom. Yet the pursuit of theory as a form of practical engagement is not unique to liberal theorists. In fact, I argue that antiliberals are equally attentive to liberalism and equally concerned with the practical effect of their criticism yet assume that only their contrary stance can enable its effectiveness. In the remainder of this chapter, I explore this quest for a form of social criticism that will be effective in so-called liberal democratic societies. For an engagement with liberalism to advance this quest, however, assumes much more than these theorists typically make explicit. To unpack some of these assumptions will require us to understand more clearly what we are talking about when we talk about "liberalism."

What do theorists understand liberalism to be? In popular discourse in the United States, at least since the late 1980s, the term "liberal" has been invoked as a term of abuse more commonly than it has been a term of approbation. This shift has also appeared in Britain. Yet at the same time, on both sides of the Atlantic, liberalism has become an ever-more widely embraced moniker among academic political theorists who use it as a descriptor for a dominant ideology that encompasses a wide spectrum of political positions, including many conservatives and social democrats in addition to so-called liberals.[7] For political theorists, the meaning of liberalism has become simultaneously more expansive and more contested. As John Gray argues,

> One of the curious features of the present time is that, even though we are all liberal, there is no agreement about what liberalism means. Some people will tell you that the core liberal value is personal liberty, but others insist it is equality. Some say that liberal values require multiculturalism, while others believe they demand a common culture based on personal autonomy. For some, liberalism is a strictly political theory that applies only to the structure of the state. For others, it is a whole way of life.[8]

Here, Gray highlights the ambiguity inherent in any claim that "we are all liberal." To make sense of this, we must unpack divergent understandings of what liberalism is. There are two key dimensions that I will pursue here: liberalism's *scope* and its *character*.

By focusing on scope, we can ask how far various conceptions understand liberalism to reach. Does it apply, in Gray's terms, "only to the structure of the state"? In this sense, being a liberal is synonymous with an embrace of familiar liberal democratic government institutions: constitutionalism, elected representation, rights-claims, church-state separation, and protection of private property. Yet when liberalism is conceptualized in this limited, institutional manner, there is no reason to expect popular attitudes or practices to be consistently liberal. In fact, one familiar (liberal) justification for these institutions is to create a legitimate space in which individuals or groups might promote policies, engage in practices, or hold opinions that might be non- or illiberal. I might adhere to a religious faith that proclaims God's absolute authority and be convinced that only true believers can escape eternal damnation. I might join a group aiming to advance collective ownership of property. I might advocate direct, participatory democracy as an alternative to representation. Yet so long as I channel my advocacy and activities within legally allowed avenues, I should expect little interference from the institutions of the liberal state. In this sense, a liberal structure alone does not necessitate or even expect that individuals living within it will *be* liberal.

But what if liberalism is much more expansive in scope? What if it is, to borrow again from Gray, "a whole way of life"? If so, then we would find that prominent practices and policies, as well as the character of social movements and civil society, are recognizably liberal. Here we would find a political culture that fosters consistently liberal citizens and liberal public opinion. This has long been an influential characterization of the United States. Theorists—including Louis Hartz, who famously argued for the liberal idea's "utter domination over the American mind"— have presented it as an exceptional characteristic of American political history and culture rather than a commonplace throughout the "west."[9] Yet in recent years, many political theorists have generalized this account, implicitly rejecting Hartz's claim of American exceptionalism. Wissenburg, from the Netherlands, characterizes liberalism as "the dominant ideology and philosophy of our times."[10] British theorist Raymond Geuss asserts that "liberal ideas permeate our social world and our everyday expectations about how people and institutions will and ought to act; they constitute the final framework within which our political thinking moves."[11] Neither of these arguments are focused upon specific national identities and instead present "our times," "our social world," and "our political thinking" as shorthand accounts of—at the least—the affluent, postindustrial societies of "the west." These claims make sense of the idea that theorists must focus upon liberalism if they are to understand these societies and to be effective social critics.

If we accept this expansive conception of liberalism's scope, we would not merely expect the state to tolerate diverse religions; we would also expect religious citizens to promote tolerance of other religious practitioners and of atheists.[12] We would expect liberal citizens to regard private property and the market as sacrosanct, to respect individual rights as inviolable, and to regard virtue or moral convictions as anathema to politics. More generally, however we understood liberal constitutional and institutional strictures we would regard these as not merely the creation of spaces for the legitimate contest of values and practices but as good in and of themselves.

The second dimension upon which conceptions differ is with regard to the character of liberalism itself. As Gray notes, this includes the relative priority of individual liberty or equality, and commonality or cultural diversity. More generally, this also encompasses the question of how rigid or malleable liberalism itself is understood to be. Writing a century ago, Eduard Bernstein argued that "society of today is no firm crystal, but an organism capable of change and constantly engaged in a process of change."[13] Here Bernstein was reappropriating Marx's own words to rebut fellow Marxists who believed that reform from within was impossible and revolutionary transformation both possible and necessary.[14] He observed that they implicitly treated capitalist society as inflexible and brittle. Bernstein sought to counter this belief; describing society as both capable of change and constantly changing allowed him to capture this notion of pliability and adaptability. Although the context was different, this contrast is also relevant for our purposes.

Is liberalism best understood as a "firm crystal"? Is it a rigid and unitary doctrine that accounts for the all the workings of liberal democratic society? If so, then critics must position themselves on the outside and shatter it, unless they wish to forgo their critical faculties and confine themselves as players within its narrow limits. As we will see, many environmentalist critics of liberalism describe it in just this way. Yet if liberalism has the character of "an organism capable of change"—more putty than crystal—then we might reasonably expect it to be stretched and adapted to meet new challenges. Liberalism understood in this way is malleable, contested, and plural.[15] Here an environmentalist might find room to criticize society from inside the liberal tradition.

In the remainder of this chapter, I analyze recent work by environmental political theorists—capturing the antiliberal environmental approach first and then contrasting that to the liberal turn. I focus less upon their evident differences than upon their shared preoccupation with liberalism and its relevance to the quest for effective social criticism (the *work* the engagement with liberalism is meant to do) and on the understandings of liberalism itself (its character and scope). In this way, I explicate some of their less apparent differences over the role of social critics: between outside critics and those who seek to criticize from within. Yet I argue that by positioning themselves as either outside or inside of *liberalism*, theorists reify its power to impede or authorize viable strategies for criticism and change.

There are unacknowledged tensions regarding these questions of scope and character in the engagement with liberalism. I point toward a more robust approach that avoids these tensions—but not by seeking to resurrect illiberal ideology or traditions. Instead, I conclude by suggesting that greater critical purchase can be found by explicating and engaging with values embedded in material practices within society. The significance of these embedded values is often hidden in discussions that treat liberalism—and other philosophical traditions—as coherent and encompassing systems that reflect both ideas and lived experience. In sum, unlike the theorists I discuss here, I aim neither to praise liberalism nor to bury it but to open a different sort of conversation that does not rely upon either one.

Engaging "Liberalism"

For environmental political theory, the engagement *with* liberalism predates the more specific turn *to* liberalism. Through the 1990s, the most prevalent sort of discussion of liberalism was arguments condemning its destructive influence. In the subsequent decade, this tendency shifted strongly toward those that sought to engage with liberalism. In this section, I evaluate both sorts of environmentalist readings of liberalism—those that reject it and those that turn to it as a means of addressing environmental concerns. I identify both their insights and limitations for a dialogue about effective social criticism.

Antiliberal Environmentalism

In a comprehensive survey of environmental thought published in 2002, Peter Hay argued that many environmental theorists had come to see "'liberalism' as the arch-opponent of its own life-privileging philosophy."[16] Two exemplary books from the 1990s that develop this position in detail are William Ophuls's *Requiem for Modern Politics* and Joel Kassiola's *The Death of Industrial Civilization*. I draw upon these works to sketch the contours of "liberalism" as understood by those who find it necessary to reject it.

William Ophuls is best known for his much-debated 1977 work, *Ecology and the Politics of Scarcity*.[17] His more recent books have garnered relatively less attention yet aim both to clarify and advance the project staked out in that initial work.[18] In his 1997 *Requiem for Modern Politics*, Ophuls argues that all modern politics is liberal politics. Because he is convinced that this politics has been severed from any moral context—Hobbes is described as the initial culprit—modern societies are unable to be *governed* in the proper sense of this word. These societies are unable to establish the sort of limitations needed for the pursuit of meaningful, fulfilling human lives and, as a consequence, we are unable to live sustainably and wisely within a finite ecological system. "We are therefore in a deep political crisis and confront the necessity for an abrupt and decisive change to a radically different type of politics," Ophuls proclaims.[19] At the core of this crisis, he argues, is "the impending failure of liberal polity, the modern system of politics founded on the tenets of classical liberalism and the rationalistic philosophy of the Enlightenment. Liberal polity is based on intrinsically self-destructive and potentially dangerous principles. It has already failed in its collectivist form and, contrary to the view of many, is now moribund in its individualist form as well."[20]

The encompassing character of this vision of liberalism becomes the justification, in turn, for Ophuls's total critique. His understanding of the characteristics of modern, liberal polities are reflected in his chapter titles, which include "moral entropy," "electronic barbarism," "predatory development," "fallacious affluence," "fraudulent abundance," "irrational reason," "intrinsic totalitarianism," and "democratic despotism."[21]

In a similar vein, Joel Kassiola has argued that "whenever . . . modern society in general is considered, the political philosophy of liberalism must be addressed."[22] Following Sheldon Wolin, Kassiola argues that liberalism has eviscerated both the scope and the significance of "the political," to be replaced by a conception of politics as "fundamentally coercive and a necessary evil to protect property and maintain law and order."[23] Thus, he characterizes liberalism as having reduced politics to market economics and liberal societies as ones that promote the seemingly "impersonal decisionmaking" of the market, which serves to obscure its politics and "disguise its actual power structure from even its own elite beneficiaries as well as from the victims of this power structure: the masses."[24] The silver lining of environmental crisis and recognition of limits to growth, Kassiola maintains, is that it "'blew the cover' of the disguise of the ideology of unlimited economic growth, [so that] we now have the first opportunity since industrialism began to assert not only the irreducibility of the political allocative process but, moreover, to insist that this process should not be obscured to protect one social group's distinct advantages."[25] In characterizing liberalism in this all-encompassing manner, both Ophuls and Kassiola link their dismay with the inadequacy of the instrumental mechanisms for environmental protection in liberal democratic states to the need to reject liberalism itself.

Perhaps the foremost point to notice for our purposes is that this critique of liberalism is premised not only upon a connection between this political philosophy and the modern political institutions but upon a deep connection to our lives within it. Andrew Vincent captures this view as well, arguing that liberalism has "entered deeply into the psyche of Western nations over the last two centuries . . . [and] constitutes a major part of our 'self-image' in Western industrialized societies, both philosophically and practically."[26] This deep connection can seem self-evident; after all, we describe these societies as "liberal democratic." If this is accurate, then isn't it a truism to conclude that liberalism is the philosophical foundation of liberal-democratic societies? The answer is no. Only if the scope of liberalism is encompassing does it become so. To posit a singular, coherent liberalism as the core of contemporary societies is—at best—an argument in need of defense rather than a self-evident fact about these societies. Yet, as I describe later in this chapter, this role for liberalism is also central to the project envisioned by many seeking to reconcile liberalism with environmentalism.

When liberalism is understood in this manner, to critique it is to critique the conceptual apparatus by which citizens of liberal democratic societies are said to understand themselves and make sense of their world. To reject liberalism is thus to contend that the dominant worldview of these societies must be transformed; that the self-understandings of its citizens must be rejected. In this sense, authors such as Kassiola and Ophuls offer a totalistic critique of liberalism, because they regard liberalism as a totalizing ideology. Meaningful inside criticism is regarded as impossible, because it is necessarily complicit with the destructive tendencies inherent in liberalism.

Just as the scope of liberalism is construed by Ophuls and Kassiola to be broad and hegemonic—the encapsulation of the public philosophy of contemporary liberal-democratic societies—its character is depicted as rigid. Privileging private preferences expressed in the marketplace, liberalism is characterized as promoting possessive individualism and unbridled consumerism. In the name of state neutrality, it precludes meaningful restraints upon the ownership or use of private property, excludes collectively defined conceptions of the good, and devalues citizen virtue or public-regarding action. All this radically constricts the public sphere of democratic politics in favor of the private sphere of economic preferences and narrowly self-interested action.

This picture of liberalism comes close to equating what some thinkers label "classical" liberalism with the character of the broader liberal philosophical tradition per se. Of course, Kassiola, Ophuls, and other antiliberal authors are aware of competing strands within liberalism—that many self-identified liberals over the past century do not subscribe to core tenets identified with classical liberalism. Yet their position is that these differences within liberalism do not, in the end, truly *make a difference*. From this perspective, it appears that the classical liberal tenets overwhelm the efforts of many more recent thinkers to modify or remove these from the liberal canon.

In sum, two premises are central. First, that citizens in liberal democratic societies can be characterized as liberal. Second, that "liberalism" mandates privileging the private at the expense of the public, even when other liberals seek to modify this stance. Liberalism, here, is a unified theory rather than a broad tradition of ideas. If liberalism requires the exclusion of the good from political life, a priority on the protection of private property rights, and an embrace of a muscular sort of possessive individualism, then it will not have the capacity to adapt or incorporate challenges to these commitments.

With these two premises, a coherent picture of liberal democratic societies emerges from the antiliberal environmentalist critique: these societies are rooted in an inflexible worldview that precludes effective criticism or change from within. It is encompassing in scope and rigid in character. In Bernstein's terms referenced earlier, liberal society is a "firm crystal." This picture is key to the authors' diagnosis of liberalism as the disease that has caused our current environmental malaise. The diagnosis then suggests that a cure for modern society can only be effective if it confronts the liberal disease directly. Unless or until that is the case, environmentalist efforts will prove to be mere bandages that temper superficial symptoms rather than curing the underlying pathology. Moreover, such efforts will likely prove counterproductive, because they can bolster support for liberalism, contributing to complacency at exactly the moment when the disease itself might otherwise become more clearly understood and attacked.

When generalized, this dilemma—that reformist efforts may not be just ineffective but counterproductive—is not a new one. Socialists have been debating it for a century and a half, as some environmental thinkers have noted.[27] The argument, at its core, is that the encompassing scope and inflexible character of the extant system is precisely what requires that it be overthrown—*and* what makes this possible.[28] Transformational change is thus the only cure for the liberal disease and so criticism from outside the liberal consensus the only type that can prepare the ground for such change. In sum, the rejection of liberalism by environmental theorists such as Ophuls and Kassiola reflects their diagnosis. In the absence of attention to the underlying power, reach, and rigidity of this system of ideas, these authors contend, environmental efforts will prove futile.

The Environmentalist Turn to Liberalism

In light of the preceding discussion, environmentalist thinkers' more recent turn to liberalism invites closer scrutiny. There are two distinct types of argument that promote a liberal-environmental rapprochement, only one of which I will closely examine here. The first is unambiguous in advocating classical liberalism. Here, libertarians and "free-market" enthusiasts turn the arguments of authors including Ophuls and Kassiola on their head. Whereas those authors described environmental externalities such as pollution or species extinction as the consequence of private interests pursued through the market, Terry Anderson and Donald Leal assert that "free market environmentalism emphasizes the importance of market processes . . . Only where rights are well defined, enforced, and transferable will self-interested individuals confront the tradeoffs inherent in a world of scarcity . . . Even externalities offer profit niches to the environmental entrepreneur who can better define and enforce property rights to the unowned resource and charge the free rider."[29] These authors propose (their interpretation of) classical liberalism as the *cure* for environmental problems, rather than the disease that caused them.[30] This market-based liberal conception has become especially prominent in recent years among legal scholars concerned with the environment.[31] Given the centrality of the concept of private property among proponents of this position, I will devote attention to it in chapter five. It is not, however, the position that I focus upon here.

A second type of environmental engagement with liberalism seeks to stretch liberalism beyond its classical or libertarian conception, with the aim of creating more space for environmentally sensitive practice. Beginning around 2000, a significant number of political theorists have pursued this engagement.[32] Of course, there are differences between these theorists. Several would be reticent to identify themselves as liberal. All, however, have sympathetically explored the question of whether a reinterpretation or reconstruction of liberalism could adequately incorporate environmentalist concerns. All offer reason to conclude that environmentalists can be liberals, so long as the character and scope of liberalism is properly understood.

A common move among these theorists is to distinguish between two distinct strands of the liberal tradition. Avner de-Shalit, John Barry, and Piers Stephens, for example, each distinguish classical liberalism from a political or social liberalism by arguing that only the former is "intrinsically, as opposed to contingently or instrumentally, tied to capitalism."[33] Although this also invites the question of how one defines capitalism, it is worth noting that this position begins by distancing itself from precisely the associations that the critics discussed previously identify with liberalism itself: a laissez-faire conception of the state's relation to markets and capitalism, its privileging of private property and possessive individualism, and the burden of proof placed upon those who seek to justify regulation of the economy.[34] Stephens argues that this social liberalism is a basis upon which one can oppose "the neo-Weberian instrumental rationality manifested in the contemporary capitalist order."[35] In these ways, environmentalist political theorists build upon the distinction between varieties of liberalism described by many others over the past century. As Marcel Wissenburg argues, the environmental critique often leveled against "economic liberalism . . . cannot reflect upon political liberalism."[36] Although classical liberalism is often identified with Locke or Nozick, the more expansive liberal vision is more likely to draw upon J. S. Mill, the pragmatism of James and Dewey, and Rawls.

As noted at the outset of the chapter, this expansive reading of liberalism has been adopted by many contemporary political theorists not focused upon environmental concerns. Ruth Abbey persuasively characterizes their motive as follows: "By approaching the liberal tradition in this more plural way, such thinkers revive subtraditions from within the history of liberal thought in order to question and counteract the currently dominant interpretations of liberalism that they find unappealing or unpersuasive . . . It is precisely this broadening of the tradition that allows thinkers to challenge key aspects of contemporary liberalism while continuing to call themselves liberals."[37]

The fact that the expanded liberalism embraced by these theorists is centered on precisely those commitments that antiliberal environmentalists, including Kassiola and Ophuls, contend are excluded is more than a curiosity. It leads to the conclusion that these two sets of authors share many of the same concerns and objects of criticism. Yet the latter interpret liberalism as a cohesive and unified philosophy—a firm crystal unable to be changed—requiring critics to position themselves outside the tightly drawn liberal boundaries. By contrast, the authors considered in this section reinterpret the character of liberalism as a broad and flexible tradition of thought—putty-like in its ability to stretch and adapt to changing interpretations and challenges. Here, the differences within the liberal tradition *do* make a difference—precisely the position denied by antiliberals. As a consequence, these authors position their criticisms inside this tradition rather than outside it.

Rethinking the centrality of an unencumbered individualism and instrumental rationality while challenging proscriptions against regulation is a position shared by those promoting an expansive liberalism. The ability to pursue such an agenda successfully obliges these thinkers to grapple with the status of proceduralism and neutrality within the liberal tradition. Yet on this status there is significant disagreement. The key question is whether or not liberals and the liberal state must remain neutral between competing conceptions of the good.

Let us begin with a familiar argument: environmental sustainability involves a normative conviction that we are protecting or sustaining something that is good. Liberalism—and the liberal state—is often taken to prioritize the right before the good and hence to entail a commitment to neutrality among conceptions of the latter. Thus Ophuls argues that "we now seem obliged to transcend the liberal paradigm. That is, we must restore the moral or spiritual context of politics, finding some way of making politics once again rest on virtue (rather than on mere self-interest) without at the same time resurrecting the intolerance and oppression associated with almost all forms of religious politics, ancient or modern."[38] Yet if de-Shalit, John Barry, and Stephens are correct, this embrace of virtue and the good need not "oblige" us to "transcend the liberal paradigm" after all. Instead, we might find room for the good within the liberal tradition by highlighting an alternate strand within it. After all, Stephens (following Jeremy Waldron) observes that the use of "neutrality" as a "defining characteristic of liberalism" is only a product of the later part of the twentieth century.[39] De-Shalit summarizes this position well:

There are two conceptions of politics which represent two interpretations of liberalism. One, which is generally speaking more common to contemporary mainstream American liberalism, is based on the value of neutrality (hence what falls under the category of the "political" is rather limited), and amounts to minimal state intervention, opposition to regulation, and a concept of politics as an aggregate of autonomous decisions—all of which are antithetical to environmental policies. The other interpretation, sometimes referred to as "social liberalism," is not hostile to advancing certain ideas of the good (e.g., conservation), and is more open to state intervention.[40]

As a result, the question moves away from the abstract one of whether "the good" is consistent with liberalism and replaces it with the more concrete question of whether the particular sort of good represented by a commitment to environmental sustainability is consistent with liberalism. These authors answer yes to this latter question.

This answer is bolstered by the argument that liberalism has always entailed a commitment to some conception of the good—that neutrality has never been more than a useful fiction for liberalism. Thus the liberal state is not in the position to choose whether or not to commit itself to the good of environmental sustainability or to remain neutral but only to choose which commitment to embrace.[41] So-called neutrality thus represents a commitment to the values that are advanced when the state withholds action or protection. This is a particular conception of the good rather than the absence of such a conception.

Others are not convinced that they need to jettison liberal neutrality. Andrew Dobson worries that the position summarized previously constitutes a "full-frontal assault on what many continue to regard as a jewel in liberalism's crown: its commitment to neutrality as far as 'comprehensive doctrines' are concerned."[42] Instead, he argues that liberal state neutrality actually favors environmental sustainability far more than previously recognized. While some embrace this position out of an abiding conviction that neutrality is a worthy value to be protected (Marcel Wissenburg is a prominent voice here), Dobson appears to do so for more instrumental reasons. Thus, he goes on to argue that "the tactical disadvantage [of arguments that reject neutrality] . . . is that they depend on many modern liberals performing a major volte-face before they can even begin to think about instantiating environmental sustainability."[43] Instead, he argues that "being serious about neutrality involves a more active and normatively self-aware state than we might suspect" because liberal freedoms require the protection of our options and opportunities to pursue our own visions of the good life.[44] Thus, the state's neutrality obliges it to protect and preserve these options and opportunities. For Wissenburg, this leads to the delineation of a "restraint principle" that provides an entrée for green concerns yet is consistent with liberal neutrality. He defines this principle in the following terms: "No good shall be destroyed unless unavoidable and unless they are replaced by perfectly identical goods; if that is physically impossible, they should be replaced by equivalent goods resembling the original as closely as possible; and if that is also impossible, a proper compensation should be provided."[45]

This offers a dramatic challenge to conventional conceptions of liberal property rights and conceptions of ownership, "thus turning ownership rights into rights to use rather than (absolutely) possess."[46] This characterization of liberal property rights as use rights—*usufruct*—"effectively converts property holders into ecological trustees with obligations to both the present and future generations."[47] As noted, Wissenburg is one of the most unabashed liberals among contemporary environmentalist thinkers. By separating the liberal concept of private property from ownership in this way, however, he poses a striking challenge to the conventionally understood character of the liberal tradition. More to the point, his restraint principle can be understood to prevent the foreclosure of options or opportunities available now to some people from being available to others in either the present or future.

Bryan Norton advances this project in an especially interesting way. Arguing that "options are a prerequisite of true freedom," Norton contrasts two strategies for protecting these options.[48] The first seeks to compare "aggregated welfare of individuals, qua individuals, across generations."[49] This abstract utility comparison is central to the discipline of welfare economics. It treats physical goods as fungible; therefore they do not play a direct role in determining of our options and opportunities for well-being. By contrast, Norton advocates a qualitative approach that specifies "stuff—aspects of the natural, physical world, which we should not allow to disappear or be degraded if we want the future to be as well off as we are."[50] He affixes the prosaic label "Listing Stuff" (LS) to this approach. This requires us to identify and describe aspects of the physical environment that are vital to ensuring opportunities and options, including "important sites, biological taxonomic groups, standing stocks of resources, and important ecological processes. Examples of stuff would include: adequate supplies of fresh, clean water; the Grand Canyon; grizzly bears (or, more generally, 'biological diversity'); an undiminished ozone shield in the upper atmosphere; and also perhaps landscape features, such as a predominantly forested landscape."[51] A key challenge for this approach, it would seem, is the process of "listing" the relevant "stuff" in the first place. Norton envisions the process as follows: "Deciding which options are most important, and most worthy of protection, becomes (in a democratic society) an exercise (hopefully participatory) in community self-definition."[52] I will return to this exercise in communal self-definition. For now, we should see that through this process certain "'indicator' characteristics" will be identified "that can be expected to correlate with the protection of these especially valued options within that environment."[53] The protection of these characteristics thus reinforces a place-based identity that can serve, in a virtuous circle, to support further efforts to protect the physical characteristics that provide valued options. In all of this, Norton seeks to respect the liberal state's goal of furthering individual freedom by ensuring that options and opportunities are not foreclosed. By characterizing the "stuff" needed as a precondition for such opportunities, it becomes a precondition for ensuring individuals have the ability to pursue a wider array of conceptions of a good life than would be possible in its absence. As Dobson argues, it "commits the liberal state to these things too, if it is not to stand accused of non-neutrality by omission."[54]

In sum, although these environmental authors offer often-rich reflections on political values, their embrace of liberalism is concomitant with a deep disagreement of what it is that they embrace. Why the turn, and why now? Although ironic—given that much of this work is carried out at a high level of abstraction—this turn is often driven by practical concerns for effective criticism and change in contemporary societies. To position the work in this way, they draw upon two related views to reinforce the focus upon liberalism.

The first is a "recognition of the fact that liberal democracy is both here to stay and has intrinsic value," a clear contrast to earlier "environmentally oriented political theorists [who] assumed that 'the' ecological crisis required . . . a kind of green revolution."[55] One might detect a note of regret mixed with the "recognition" of liberal democracy's staying power. Embedded in Barry and Wissenburg's statement here is a conflation of two divergent conceptions of the scope of liberalism. On the one hand is a claim about the stability of "liberal-democratic" political institutions; on the other is a claim that these consistently embody liberal-philosophical ideas. This conflation was also apparent in the arguments of the antiliberals discussed in the previous section.

The second view that reinforces the turn to liberalism posits that public opinion and values in western societies consistently reflect liberal norms. When this is assumed to be true, the debate about liberalism can be treated as consequential for environmentalists' criticisms and their efficacy in promoting social change. Here, the differing interpretations of the scope of liberal theory I have described serve as a surrogate for disagreement about whether environmental concerns resonate with existing public attitudes. To embrace liberalism is to advance a form of environmental criticism designed to engage extant public values, whereas to reject liberalism is to suggest that these values are inherently problematic for environmentally beneficial change.

To use a disease analogy: the symptoms of environmental destruction that worry Kassiola, Ophuls, and others who reject liberalism also worry those who embrace and rework liberalism, such as de-Shalit, Norton, and Wissenburg. Thus, the turn to liberalism need not reflect a greater complacency toward, nor a different conception of, environmental concerns. What about the diagnosis, however? Here again there are many important—and surprising—similarities. Like the antiliberals, these liberal environmentalists blame the resistance to regulation and restraint often found within existing liberal democratic polities, the privileging of unconditional conceptions of private ownership, and the depoliticizing effects of either state neutrality or the failure to recognize the biophysical preconditions for the individual freedoms that state neutrality ought to ensure. Key to their diagnosis is a strategy that first expands the interpretations of the liberal tradition and then proposes to cure our present environmental malaise by moving us from one conception of the character of liberalism to another—the latter under-recognized within this tradition.

Thus the diagnosis offered by liberal environmental thinkers is more similar to than different from that of antiliberals. The conflict that the latter identify between liberalism and its critics is now relocated within the liberal tradition itself. The most evident difference involves the relabeling of the diagnosis—not liberalism itself but a particular strand of liberalism—as the cause of the disease.

Environmental Criticism and Everyday Practice

Although I have been describing these different diagnoses as the reflection of distinct liberal and antiliberal schools of thought, one can locate both of these moments in Andrew Dobson's book *Citizenship and the Environment*. For Dobson, an appeal to citizenship stands in contrast to the strategy of imposing taxes and incentives to structure self-interested behavior; it encourages us to attend to the role of education and socialization in fostering the virtues necessary for action and so provides a more deeply rooted basis for ecologically responsible "habits and practices."[56] Citizenship also serves as a bridge between the individualistic character of moral concern that is prominent in the literature of environmental ethics, and the need for engaged, collective action to respond to environmental challenges.[57] Central to this analysis is his distinction between *"environmental* citizenship" and *"ecological* citizenship." Whereas the former reflects "a liberal point of view," the latter is "postcosmopolitan."[58] Postcosmopolitan citizenship, according to Dobson, contrasts sharply with key tenets of both liberalism and civic republicanism. It stresses the following:

Responsibilities rather than rights, and . . . regard[s] these responsibilities as non-reciprocal rather than contractual, thus contrasting with both liberal and civic republican understandings of citizenship obligations. It also focuses on virtue as being central to citizenship, but once again differs markedly from both liberal and civic republican articulation in its sense that these virtues need to be drawn from the private as well as the public arenas. Similarly it unusually regards the private arena as a legitimate site of citizenship activity, both because the kinds of relationship normally associated with that arena are similar in content to those of ecological citizenship, and because the private realm generates the space—the ecological footprint—that gives rise to the obligations of ecological citizenship itself.[59]

Here, Dobson argues powerfully that liberalism cannot offer a public philosophy able to foster ecological citizenship. In doing so, he aligns his critique with central points articulated by Kassiola and Ophuls. It seems clear at this point in the book that Dobson is urging the rejection or transformation of liberalism as necessary for enacting ecological citizenship.

With this in mind, it is bracing to find Dobson asking—on the very next page—"how . . . education systems in liberal states [can] teach ecological citizenship" and devoting the remainder of his book to a reflection "on the place of ecological citizenship in liberal societies."[60] He uses "liberal societies" as shorthand for the existing nation-states of North America, Europe, Australasia, and others.[61] He wants to do more than theorize citizenship; he seeks to engage actual citizens. Thus, despite leveling a systematic critique of liberalism (from the outside) in the first half of his book, his equation of existing societies with liberal ones seemingly requires him to retreat within liberalism if he is to promote citizen action.

Although Dobson's rejection and reengagement with liberalism is unusual within a single work, the strategic calculation represented by the return is not. It embodies the combination of resignation and engagement that is characteristic of much work that pursues a liberal-environmental rapprochement. It is a stance that we might term that of the chastened social critic, one now intent upon working with "the people" rather than advocating radical views outside their lived experience. It reflects a perceived divide between intellectually coherent yet "utopian" critique on the one hand and politically engaged, immanent critique on the other.[62] The transparent character of Dobson's move from the former to the latter makes it evident, in a manner in which it often is not, that the basis for the inside critic's focus upon liberalism can be the conviction that the latter is the public philosophy of the citizenry of western nation-states.

Why, however, should we presume that citizens in postindustrial societies hold views that can be consistently identified with the canon of liberal philosophy? Recognizing this as a presumption—not a fact—allows us to open up and explore a space for criticism immanent to these societies without first requiring it to be immanent to something labeled the philosophy of liberalism. To be clear, given the divergent conceptions of liberalism available, one could often make the claim that immanent social criticism is consistent with, or within, a particular one of these conceptions. Yet doing so invites counter claims and adds little to the resonance or relevance of the initial criticism. Better to foster environmental social criticism that engages with public concerns and values directly and bypass the contest over conceptions of the character and scope of liberal theory—which simply cannot be a surrogate for engaging public concerns.

Why is it so hard to promote change? Liberal environmental arguments suggest that it is because environmental concerns have not previously been couched in explicitly liberal terms. In this sense, they hope that their criticisms will now resonate with popular views. Thus they hope to avoid being dismissed as utopians tilting at windmills or perceived as elitists who disdain widely held values as false consciousness. By situating this form of criticism entirely within the liberal tradition, however, they usurp our ability to identify popular values that *cannot* readily be discovered in liberal philosophy, foreclosing a political space that—although offering no guarantees or easy answers—can be a crucial location for constructive environmental criticism. Opening up this space for ambitious yet immanent criticism can enable us to think in fresh ways about remedies for environmental problems. It allows us to escape from—or perhaps burrow in between—an unproductive dichotomy between the transcendent and totalizing social criticism practiced by antiliberals and the circumscribed and resigned form often practiced by those who have embraced liberalism.

Throughout this chapter, I have surveyed arguments about liberalism by theorists who imagine it as having an expansive scope, noting the traps and presumptions it leads to in their work. If, however, we return to a much more limited conception of what liberalism is, one narrowly focused in the way John Gray—in the passage quoted at the beginning of this chapter—described as "a strictly political theory that applies only to the structure of the state," then we might delink the task of environmental social criticism from much of the debate about liberalism. This is consistent with Robyn Eckersley's characterization of a green democracy that is "postliberal rather than antiliberal" in the sense that it takes leave of debates about liberalism yet need not abandon "the enduring features of the liberal democratic state, such as the protection of civil and political rights . . . the election of parliamentary representatives, the separation of powers, the idea that state power should not be absolute or arbitrary . . . and the idea of toleration."[63]

This narrowly conceived scope of liberalism isn't exactly new to environmental thinking, though it has been neglected by many in recent years. In the 1980s, philosopher Mark Sagoff was one of the earliest to suggest a liberal-environmental rapprochement, and did so on just this terrain. In a chapter entitled "Can Environmentalists Be Liberals?," Sagoff rejects the notion that liberal theory can or should delineate a "comprehensive view" from which policy is derived.[64] If theory was comprehensive, a liberal commitment to neutrality would severely restrain both the scope of state action and the character of policy. Yet Sagoff argues that such an encompassing scope of liberalism would be inherently antidemocratic, because it obstructs the ability of the people to advance the good through political action. He contrasts this with a far more modest conception of liberalism as only delineating the basic structure of the polity and protecting a minimal private sphere for religious activity and intimate life. Sagoff argues that there is nothing in this private sphere that poses a challenge for environmental policy; "I cannot think of any environmental statute that restricts . . . personal choices or beliefs" regarding friends, religion, or sexual relationships.[65] Rather than seeking the expansion of liberalism to encompass environmental concerns, then, he argues for a far more limited scope of liberalism that does not measure environmentalist commitments or motivations for political action on a liberal yardstick. In this sense, environmental action is outside the liberal versus antiliberal debate, and thus advocates have little need to focus upon it.

Here, *democracy* is not only allowed but cultivated in civil society and policymaking, in which particular conceptions of the good necessarily come into play. As Sagoff describes it, this allows for values that members of a community share as citizens to trump private-regarding preferences they hold as consumers. To preclude these citizen values, in Sagoff's view, is to commit a category mistake, by presuming that liberalism required the privileging of private values over the public ones that individuals also, in fact, hold.[66] The key liberal qualities to be fostered in this context would be a sort of nondogmatic liberality, open-mindedness, and critical thinking that facilitates an open and vigorous policy debate—rather than any commitment to neutrality or other inflexible impediment to environmental policymaking.[67] In the course of defending liberalism, then, Sagoff imagines a shrunken scope for its exercise. It is not liberal philosophy that captures the character of popular values and concerns, but it is the latter that are crucial to advancing the pursuit of sustainability.

To pursue this alternate project requires us to resituate our critical engagement from one within liberal philosophy to one within the practices we hope to influence. In one sense, then, liberalism is too narrow a field for environmental criticism—because popular attitudes regularly spill across its boundaries. At the same time, it can be too broad a field, because any particular conception of liberal ideas might transcend those embedded within contemporary practices and values. Environmental challenges are prompted or exacerbated by practices of work, play, consumption, and ownership. These practices are, to say the least, complex. By engaging directly with this complexity, we can tease out and highlight those aspects that we judge to hold promise.[68]

In this manner, we can honor the impulse to develop criticism that can resonate with citizens, which I argue has fed the liberal turn, while also taking seriously the antiliberal's conviction that environmental crisis demands that we reject conceptual complacency. To engage theory with practice in this way promises to open up new democratic political space— about which I say more in subsequent chapters. This political space can best be envisioned if we minimize the question of liberalism's character to environmental social criticism. Then we can more effectively pursue a popular engagement of environmental concerns with extant or nascent public values, because we will neither allow a conception of liberalism to constrain this engagement nor allow a rejection of public values to alienate us from the broader constituency for social and environmental action.

3

The Question of Materiality in Environmental Politics

In the previous chapter, I argued for taking leave from debates about abstract philosophical traditions (liberalism in particular) and turning instead to the role of everyday practices. These practices are material ones in the sense that human actions are inescapably entwined with a larger web of life forms (human and nonhuman), natural formations, technologies, and built environments. In this chapter, I grapple with what it might mean to make this turn, examining the role that materiality plays—and ought to play—in contemporary efforts to promote environmental sustainability.

I choose to use the word "materiality" here to leave some openness regarding this role. By contrast, "materialism" and "materialist" are words that typically express a particular position regarding materiality—though what that position is depends upon the context. In everyday conversation, to say that someone is materialist—or materialistic—usually indicates a particular preoccupation with consumer goods and other "stuff." For philosophers, materialism is an account of physical matter as the whole of reality in contrast to idealism's focus on human subjectivity as something essential and apart from matter. In certain parts of the academy and in certain historical moments, to be a materialist has been also to embrace a Marxist interpretation of economic production as the central material experience, key to shaping social, cultural, and political ideas. All of these have something to do with the way that materiality relates to sustainability, but none fully captures it.

The role and position of materiality is—perhaps surprisingly—deeply divisive among both scholars and activists concerned with environmental sustainability. The divide is, at its core, between an account that minimizes the role of material conditions and practices—in favor of the centrality of vision and values—and an account that reverses this relationship. In this chapter, I will argue that this divide is itself problematic. Before doing so, however, we need a clearer sense of each of these accounts. Both can be observed when considering two questions at the heart of the resonance dilemma facing these efforts and movements.

The first is the question of environmentalist *identity*. What sorts of people are likely to be, or become, environmentally concerned? Is environmentalism—as many have claimed—necessarily a movement of educated, privileged, relatively affluent citizens? If so, of course, the movement's constituency is limited. Its relative prominence in North America, Europe, and other postindustrial societies could be explained by the relatively large size of this group in these societies. Because this socioeconomic strata is much smaller in the nation-states of the Global South, we would expect such an environmentalist identity to be far less prominent in these societies. By contrast, if environmental concern can be understood as a less elitist and more inclusive identity (or, set of identities)—if "environmental" concerns can be embraced by more diverse constituencies—then these concerns might become far more powerful and salient across classes, cultures, and positions in the world economy.

The second—related—question, then, centers on the definition of environmental *concerns*. Are environmental problems essentially, or paradigmatically, about such things as the protection of pristine wilderness and endangered species? Are they necessarily or primarily motivated by "ecocentric" values? If so, then concerns that often emerge closer to home—exposure to toxins in already "degraded" urban environments or ensuring a sustainable rate of timber harvesting from a working forest, for example—will appear less pressing or central to the environmentalist project. In such cases, the relationship between environmental concerns and the problems of everyday life and livelihood become much more central. Both questions are prominent in discussions of the relationship of environmentalism and environmental justice,[1] northern and southern movements,[2] and discussions of strategies for effective climate change communication.[3]

We can develop insight into these discussions by recognizing that the definition of both environmentalist identity and concern are fundamentally questions about the role of materiality in environmental politics. I begin with a summary of influential accounts of this role, focusing first upon the highly influential "postmaterialist values" thesis—whose most well-known academic exponent is Ronald Inglehart—as it applies to environmentalism.[4] The thesis examines postmaterialism (and so materialism) from an empirical and social psychological perspective. That is, here the labels materialist and postmaterialist are used to describe the existing views held by people in contemporary societies and to identify the social forces that shape or create these views. The prominence of the postmaterialist account of environmentalism can be explained by the fact that it appears to illuminate many facets of contemporary environmentalism in affluent societies. Yet the further one digs into this account, the more elusive and contradictory it becomes. I argue that both the direct and indirect influences of the postmaterialist account have fostered a highly distorted understanding of environmentalist identity and concerns, resulting in a shrunken imagination that has occluded meaningful strategies for change.

The question, then, is what an alternative account of environmentalism might look like. This question can only be engaged seriously after we first step out of the shadow of the postmaterialist thesis, which has not only been influential in its own right but also served to frame the other side of this coin—"materialism"—in a misleading and problematic way. I then navigate through some of the insights and limitations of recent writings by "new materialist" scholars to reconstruct the role of materiality in politics. I argue that the question is not just *whether* environmentalism is and should be connected to materiality—the answer to this question is yes—but *how* materiality matters in this context.

Materiality in Environmental Politics

"Postmaterialism" as an Explanation for the Rise of Environmental Activism

Many histories of the rise of environmentalism in postindustrial societies have been written, and of course there are important differences in the stories told. Although there is no need to rehash these stories here, I begin by offering a brief and familiar potted history of the precursors and rise of environmentalism in the United States in order to illustrate a relevant common theme: environmentalists often have been defined by concerns

that are at a considerable distance—both spatially and experientially—
from their own everyday material conditions and livelihoods. A consequence has been to elevate experts and elites to a privileged role within
environmental movements.

The oft-told story of US environmentalism traces its "prehistory" back
to the struggles of—and between—John Muir's effort to preserve certain
areas of distinctive natural grandeur and Gifford Pinchot's effort to con-
serve natural resources through their efficient use. In the case of the latter,
the reliance upon technocratic expertise was central.[5] In both cases, this
story focused our attention on places far from the nation's population
centers while the primary constituencies for these movements were
found in these centers, largely among the relatively affluent and
well-educated.

For those whose lives and livelihoods *were* directly affected by pres-
ervationist and conservationist programs—especially American Indians
and others in rural communities dependent upon resource extraction—
resistance was commonplace, and these protoenvironmental policies
were often viewed as imposing the values and interests of distant urban
elites.[6]

Much changed with the emergence of a self-identified environmental
movement in the 1960s and 1970s. Yet the constituency continued to be
composed disproportionately of materially comfortable and educated
urban and suburban residents, whereas the reliance upon expertise—
scientists, lawyers, and lobbyists—was far more central in newer environmental organizations, such as the Natural Resources Defense Council
and the Environmental Defense Fund, than it was in longer-standing
organizations, such as the Sierra Club. Moreover, although local issues
of air and water pollution came to the fore, the preoccupation with
distant wilderness remained paradigmatic of environmental concerns. In
the words of William Cronon, "although wilderness may today seem to
be just one environmental concern among many, it in fact serves as the
foundation for a long list of other concerns that on their face seem quite
remote from it."[7]

What explains the growth of environmentalism and other so-called
new social movements that rose to prominence in this era? Beginning in
the 1970s, a number of scholars posed this question. The answer seemed
unlikely to be a uniquely American one, as a similar change was taking

place in other affluent western societies, including Canada, Western Europe, Australia, and New Zealand. Much of what seemed new to scholars had to do with the distinctive social values—in particular those of autonomy and identity—that these movements appeared to privilege.[8] This sense that new movements could be explained by their focus on *values* was often explicitly contrasted to materialist explanations—whether the latter focused on biophysical conditions of the environment or on the "old politics" of labor, with its agenda of "economic growth, distribution, and security."[9]

Political scientist Ronald Inglehart offered one of the most enduring explanations of the role of social values in these movements, values that he has sought to measure through decades of ambitious, cross-national, public-opinion surveys. It's not so much the survey data itself but the theoretical framework that Inglehart adopted to interpret it that merits our attention here. Inglehart analyzed his data through the lens of psychologist Abraham Maslow's well-known hierarchy of human needs. Maslow had described humans as motivated by the desire to fulfill a ranked order of needs, beginning with the physiological and then moving upward to needs for safety, love, self-esteem, and finally self-actualization. Only once the needs at one level are met do we move "up" to fulfill needs at the next level.[10] Inglehart simplified Maslow's hierarchy into a dualism and used this to explain changing values and movements in the affluent societies of the west.

These societies, in Inglehart's account, have been shifting from a materialism "driven by concern for meeting survival needs" to a "postmaterialism" characterized by an emphasis on "quality of life in general and the physical environment in particular."[11] Reliant upon Maslow's theory, he retains the notion of a hierarchy in which individuals are only motivated to act on higher-order, "postmaterial" needs once the lower-order ones are secured. Inglehart has argued that this value shift in western societies is tied to the childhood socialization of post–World War II generational cohorts in relatively secure economic conditions. He summarizes his thesis as follows: "Throughout most of history, the threat of severe economic deprivation or even starvation had been a crucial concern for most people. But the unprecedented degree of economic security experienced by the postwar generation in most industrial societies was leading to a gradual shift from 'Materialist' values toward

'Postmaterialist' priorities."[12] This dichotomy is thus consistent with, and also seems to make sense of, those protoenvironmental movements in which these concerns also emerged at a distance from concerns with livelihood. It also seems consistent with the observation that identification with the movements emerged largely among a relatively privileged and secure constituency.

In trying to make sense of this postmaterialist thesis, it is important to notice that the term "postmaterialism" is defined largely by what it is *not*. Postmaterialism is more ambiguous about what it *is*. Quality of life and self-actualization are difficult concepts to pin down, but for Inglehart these must be nonmaterialist concerns in the particular sense that they are not focused on sustenance, shelter, or the provision of other physiological needs. Inglehart regards these as ideas or beliefs rather than physical, material conditions. In this sense, the dichotomy can also be understood as between materialism and idealism (in the philosophical sense), between the concrete and abstract, and—as his later work makes explicit—between "objective problems" and "subjective values."

Although avowedly descriptive, the postmaterialist thesis also tracks with other familiar but less academic narratives about environmentalists as relatively privileged yet selfless people whose concerns are shaped by their awareness and understanding of world affairs rather than their own material self-interest. From this perspective, ignorance, greed, and egoism represent the bases of opposition to environmental protection efforts, whereas knowledge and virtue appear to lie at the heart of environmental concern. The rise of postmaterialist values in the west is often taken to illuminate the necessary precondition for environmental activism in these societies. In other words, Inglehart's theory seems to offer an explanation for why materially comfortable citizens would act to protect nonhuman species and distant landscapes, and his survey data makes it clear that there have been a rapidly growing number of such citizens in post–World War II western societies. Moreover, the theory can be used to explain why this activism is resisted by others whose material survival or livelihood is perceived to be threatened by such protections.

A similar relationship between societal affluence and the rise of environmental concern has been posited by some economists and policy analysts, termed an "environmental Kuznets curve." The hypothesis is that impoverished societies will tackle environmental problems only once they achieve a significant level of material affluence—that is, first they'll get rich, then they'll clean up. Yet careful studies have concluded that this relationship is, at best, contingent and uncertain, not a reliable basis for predicting the onset of environmental concern.[13]

Ted Nordhaus and Michael Shellenberger—activists critical of present-day environmentalism and seeking to transform the movement—also illustrate the uses to which the postmaterialist values thesis is put. In their much-discussed book *Break Through: From the Death of Environmentalism to the Politics of Possibility,* they argue with Inglehart that "around the world there is a very strong association between prosperity and environmental values."[14] They also adopt a position (which even Inglehart has backed away from, as we shall see in the next section) that "ecological concern is a postmaterialist value that becomes widespread and strongly felt—and thus politically actionable—only in postscarcity societies" and that contemporary environmentalists misunderstand their movement "as a reaction to industrialization rather than a product of it."[15] Yet while Nordhaus and Shellenberger claim that they are rejecting the conventional wisdom, in fact the postmaterialist thesis has been a prominent theme in accounts of the rise of environmental concern for at least the past couple generations.[16]

Challenges to the Postmaterialist Conception of Environmentalism

Where the postmaterialist thesis would clearly seem to offer far less insight, however, is in making sense of relatively disadvantaged citizens of these same societies or the poor citizens of societies in the Global South—those most likely to be "materialists," in Inglehart's terms—who also support, and engage in, activism on behalf of environmental concerns. By the 1990s, evidence of such support and activism was diverse and widespread. It was especially manifest in the environmental justice movement, a label first applied to those organizing in relatively poor and often minority communities in the United States[17] and later among poor people in the Global South whose livelihood was dependent upon access to natural resources.[18]

Newer interpretations and data that emerged in that decade also challenged the view of environmentalism—both past and present—as a consistent product of postmaterialism. Robert Gottlieb's revisionist history of US environmentalism, *Forcing the Spring,* challenged conventional accounts, such as that sketched previously, by identifying urban and public health initiatives of the early twentieth century as integral yet unrecognized foundations of contemporary environmentalism.[19] More encompassing was the "Health of the Planet" public-opinion survey in twenty-four nations, released in 1993.[20] As sociologist Steven Brechin summarizes its impact, "the results showed clearly that . . . environmental concern was a global phenomenon, not geographically limited, nor generated simply by the shift in values resulting from decades of economic prosperity and political security."[21] More recent public opinion research consistently reinforces this conclusion, debunking accounts that characterize the whole of environmentalism as motivated by "postmaterialism."[22]

Clearly, something more, or other, than a values shift consequent of economic affluence is at work. A different sense of environmentalism often appears here, one rooted in protection of livelihood and community. These movements and expressions of support don't just represent a new constituency for environmentalism, and so a different sense of environmentalist identity, but also offer a distinct interpretation of *what counts* as a pressing environmental concern.

Some have responded to these manifestations by denying their centrality to "environmentalism" per se. In this vein, for example, political theorist Andrew Dobson once asserted that "'traditional' environmentalism *just is* about wilderness, resource-conservation, and so on . . . it is too much to expect that these programmes will be (or should be) synonymous [with environmental justice]."[23] Indeed, some environmental justice activists have mirrored this point, concluding that they aren't "real environmentalists," based upon their own narrow conception of environmentalist identity that doesn't include them.[24] All of this should make plain that there is no singular, determinate meaning of "the" environment to be protected; diverse movements interpret this differently.[25]

There is more than one "environmentalism," with some manifestations more obviously tied to direct, experiential threats to livelihood and conditions than others. In reaction to—but still under the influence of—the postmaterialist thesis, some who study these movements have presented them as contrary, "materialist" expressions of environmentalism. Yet viewed through the lens of postmaterialism's dichotomous thesis, a materialist movement can only be understood—problematically—as one in which ideas or values either do not play a role at all or where they are characterized as determined by material experience.[26] Before illustrating the distorting effect of this dichotomy, it is valuable to consider an attempt by Inglehart to clarify or modify his thesis in the face of growing evidence from some of the challenges outlined here.

"Objective Problems" versus "Subjective Values"

By the mid-1990s, even Inglehart sought to address the difficulties posed for his thesis by these expressions of environmental concern.[27] He referenced his extensive cross-national public opinion data to acknowledge that "the relationship between values and environmental concern varies widely across nations."[28] He then asserted that there are two discrete manifestations of environmental concern: one based on the "subjective values" of postmaterialism and the other based in "objective problems" most evident in poor and more "severely polluted" countries.[29] This seemingly new dichotomy attempts to rescue at least a portion of his postmaterialist values thesis. The thesis, as applied to environmentalism, now maintains that subjective postmaterialist values explain why *some* people—that is, those in affluent societies and who are supposedly *not* directly confronted with "objective problems"—are nonetheless motivated to act.

On the one hand, Inglehart's dichotomy captures something that many observers of environmental activism might find intuitively plausible and easy to accept: in some cases public concern and support is generated by concrete existential threats ("objective problems"), whereas in other cases it is generated by abstract beliefs or ideals ("subjective values"). Superficially, this characterization appears to bring the movements and much of the data, analysis, and criticism that had previously appeared problematic for "postmaterialist" accounts of environmentalism within the fold of the theory. That is, when earlier characterizations of the post-materialist values thesis focused exclusively on "subjective values" as an explanation for public support (or its absence) for environmentalism, attention to activism or polling data showing environmental concern among the poor and disenfranchised posed a powerful challenge to the accuracy of this explanation. If this concern can be described as a phenomenon separate from values-based manifestations of environmentalism and attributed instead to "objective problems," then at least some of the explanatory significance of the postmaterialist values thesis would be preserved.

Yet although the terms are newly introduced, the dichotomy between objective problems and subjective values is not new to Inglehart's thesis. As I have argued, it is embedded within the original dichotomy between materialism and postmaterialism itself. Here, materialism is defined by its focus on "objective," existential "problems" of physiological survival, sustenance, and overcoming of scarcity, whereas postmaterialism is the term used to describe a "values"-based concern with autonomy, quality of life, and human "subjectivity." Thus Inglehart has not truly adapted his theoretical framework to new empirical data and accounts but has simply concluded that some expressions of environmentalism fall neatly on one side of the post/materialist divide, whereas other expressions fall on the opposing side. Rather than salvaging the postmaterialist thesis with regard to environmentalism, this undermines the promised explanatory power of the thesis while clinging to a dichotomy that occludes rather than illuminates.[30]

Although the identity of mainstream environmental group members in the United States and other postindustrial societies *is* skewed by class and education level in a way that seems consistent with the postmaterialist thesis, other data does not indicate the same bias. Inglehart's dichotomy between objective problems and subjective values would lead us to expect that for supposedly more abstract issues, such as climate change and other global concerns, postmaterialist values would remain decisive in explaining beliefs and actions. Yet, again, the data doesn't bear this out. Summarizing data from eighteen nations, Angela Mertig and Riley Dunlap find "that demographic variables . . . are poor predictors . . . of support for environmentalism."[31] Similarly, in his examination of global survey data, Steven Brechin concludes:

The number of postmaterialists appears to make no difference on how citizens of countries, rich or poor, respond to the more global concerns. The environmental concerns of the poorer countries appear to be based upon a broader set of values and effects than those generated simply from direct experiences. Citizens from rich countries and poor ones, regardless of the number of postmaterialists, appear to have subjective values influencing their concerns about the environment.[32]

This undermines the supposed dichotomy between (objective) problems and (subjective) values. In fact, all the overlapping dichotomies embedded in Inglehart's explanation are deeply problematic: object/ subject, fact/value, concrete/abstract, local/global, and materialism/post-materialism itself. The point is not to reject any such distinctions as being without merit. It is, instead, to recognize that when regarded as naturally given dichotomies we often fail to imagine promising approaches that rely upon the interplay of facts and values, local and global, or concrete and abstract. Yet such interplay is key, and imaginative new approaches require us to escape the straightjacket that these dichotomies have kept us in.

Everyday Environmentalism through the Lens of the Postmaterialist Thesis

It has been worth paying close attention to the postmaterialist thesis— despite the misleading answers it offers—because the question of mate-riality that it draws us toward is central to the dilemmas of environmental politics. Part of the confusion rests upon an ambiguity regarding the meaning and implications of "materialism." On one hand, when Ingle-hart writes of materialism as focused on survival and avoiding starvation the primordial nature of these needs appears to place them beneath the level on which culture and values emerge. "Objective problems," such as those now defined by Inglehart as generating environmental concern in the Global South, are characterized as threatening survival in this way. On the other hand, where "subjective values" are said to explain envi-ronmental concern—especially in the affluent societies of North America, Europe, and Australasia—the connection to material conditions is sup-posedly severed.

Yet in practice such an account is implausible. Consider a seemingly extreme case: yes, there's a minimum caloric intake that is essential for human survival, and the question of which organic substances can provide nutrients for the human body can be answered without regard to subjectivity. Nonetheless, the question of what qualifies as food (beef for Hindus? pork for observant Muslims or Jews? insects for most west-erners?) in a given context is largely culturally constructed. As Michael Walzer once observed:

A single necessary good, and one that is always necessary—food, for example— carries different meanings in different places. Bread is the staff of life, the body of Christ, the symbol of the Sabbath, the means of hospitality, and so on. Conceivably, there is a limited sense in which the first of these is primary . . . but . . . we can't be sure. If the religious uses of bread were to conflict with its nutritional uses—if the gods demanded that bread be baked and burned rather than eaten—it is by no means clear which use would be primary.[33]

If we can't neatly divorce culture, subjectivity, and values from "materialism" even in the case of food, in which the matter is so clearly necessary for biological survival, then the firm boundary between "objective problems" and "subjective values" cannot stand. The complexity of such contests and construction are magnified when we go beyond a consideration of food to that of shelter, mobility, security, and other phenomena.

The role of caregiving provides an illustration of this complexity. For some early cultural feminists and ecofeminists, a biologically essentialist argument conflated women's sex roles with gendered roles as mothers and caregivers.[34] Rejecting this view, materialist ecofeminist theorists, including Ariel Salleh and Mary Mellor, have developed an argument for the integral yet contingent role of care that does not rely upon this essentialist conflation.[35] Instead they focus upon the daily challenges, obligations, and embodied, lived experiences of caregiving that rest disproportionately upon the shoulders of a great many women. These gendered practices of typically unpaid labor are integral to sustaining life and reproducing communities. In this sense, it seems that they strive to give subjectivity its due within a materialist framework focused upon what might otherwise be termed "objective problems." Yet as significant as this corrective is, and as invaluable as the recognition of these roles and experiences in women's lives worldwide is, the introduction of subjectivity appears to only go so far. The attention to a material, caregiving role is still characterized as producing a particular perspective that appears fixed and largely univocal—that is, a perspective on politics that is ultimately portrayed by these authors as apolitical.

This avowed apolitical perspective can appear plausible because, as Sherilyn MacGregor observes, it is reinforced by a popular perception that "'motherhood issues' are not political issues."[36] The perception is both politically dangerous and conceptually problematic, because, again, it treats these evidently material roles as devoid of subjectivity and values: "the process whereby women look critically at their lives and question accepted norms . . . is necessarily diminished in ecofeminism if the assumption is that political and ecological awareness emerge intuitively (or 'naturally') from women's social location."[37] MacGregor's insight here draws our attention to the way in which the postmaterialist's dichotomy between subjective values and objective material conditions can have a pervasive and problematic influence even on those—such as materialist ecofeminists—who seek to offer an alternative.[38]

In the remainder of this section, I illustrate misrecognitions that result when the postmaterialist dichotomy is applied to contemporary environmental movements and ideas by focusing on one book mentioned earlier, *Break Through* by US political activists Ted Nordhaus and Michael Shellenberger. Nordhaus and Shellenberger first rose to prominence when they published the provocative essay "The Death of Environmentalism."[39] In both works, they lay bare a number of pressing challenges facing the movement and argue for a needed transformation of contemporary environmentalism. I focus on this book precisely because—despite what I argue are the authors' contradictions—their sustained attention to the question of materiality is a valuable basis for critical engagement.

Nordhaus and Shellenberger write that they envision a newly expansive and inclusive movement that has something meaningful to offer Brazilian workers and US professionals; that confronts "the primary public health issues confronting communities of color" in the United States;[40] and that brings together labor unions, mainstream environmental groups, social justice and religious organizations, and businesses into cross-class coalitions, such as the Apollo Alliance—which they helped found.[41] They identify the reification of "Nature" within the environmental movement as a constraint upon our ability to imagine an expansive environmental agenda. Rejecting the essentialism of a singular, capital "N" Nature, they argue for "the multiplicity of human and nonhuman natures."[42] They argue that the concerns that motivate environmentalists must transcend the familiar and conventional ones in order to connect with the salient, quotidian preoccupations with livelihood, family, health, and well-being. As they conclude, "it is hard to imagine creating a politics powerful enough to transform the global energy economy that is not fundamentally grounded in people's lives."[43] On these points, Nordhaus and Shellenberger's positions are clear and compelling.

Central to their argument is the relationship of environmentalism to everyday life, highlighting the vital question of materiality that I address here. In their book, they seek not only to advance particular criticisms of the environmental movement, but—as I noted previously—to ground these criticisms in a theoretical framework that is deeply and explicitly indebted to the postmaterialist values thesis. The result of this latter move is an account of materiality that is confused and confusing. Because the question of materiality is vital, however, it is worth taking a closer look at the results.

The postmaterialist thesis does considerable work for them. They introduce it to reject a premise that they claim is dominant: that environmental concern is the direct consequence of objective, material environmental problems. To illustrate this, they deconstruct a familiar tale in which a 1969 fire on the Cuyahoga River in Cleveland, Ohio, shocked the nation and thereby helped spark the contemporary US environmental movement. In fact, far from the first or most disastrous such fire, Nordhaus and Shellenberger rightly point out that factory effluent in urban rivers had regularly led to large-scale fires (including on the Cuyahoga) yet had long been regarded as relatively unexceptional and frequently garnered little notice in the news media. Why then was this particular fire credited with prompting environmental action? They show that an account limited to the physical conditions of the river and fire cannot explain this and turn instead to an account rooted in the changing culture and values of 1960s America.[44]

In contrast to this essentialism of material conditions, which they tie to claims for the objective authority of Science, Nature, and Reason, they rely on the postmaterialist thesis to argue that environmental action "depends on our beliefs, identities, sociocultural positions, traditions, cultures, values, and moment in time."[45] They then entwine this characterization of the postmaterialism of the contemporary United States with a central claim of their book: that the fundamental challenge facing environmentalists today is not a matter of promoting action in response to objective *facts* about environmental degradation and biophysical limits but is rooted in a political contest over *values* and visions of the good. The problem, as they see it, is that the dominant form of political engagement by environmentalists falsely represents itself as apolitical and objective.

This claim is important. However, their adherence to the postmaterialist thesis—positing objective facts as wholly separate from subjective values—leads them astray. They wrongly conflate the attention to values with a distinctly postmaterialist politics: one denatured and isolated from material conditions. In sum, their initial concern to ground their argument in people's everyday lives rightly leads them to criticize the technocratic and depoliticizing character of much contemporary environmentalism. Yet in their effort to imagine an alternative that can take seriously the centrality of identities, culture, values, and ultimately politics itself, they are drawn into the embrace of a deeply flawed postmaterialist framework that leads them to attack those approaches and movements—such as environmental justice—that have worked the hardest to cultivate the connections with everyday lives and livelihoods.[46]

The divide between materialism and postmaterialism might be conceived in one of two ways. First, materialism could truly represent a bare physiological threshold for life based upon the satisfaction of minimal caloric intake, shelter adequate to prevent death by exposure, and so forth. Living just at or below the margins of subsistence, we might imagine that *no* meaningful politics would be able to emerge, nor would the opportunity to develop or articulate values.[47] If this is materialism, then it seems coherent to associate "postmaterialism" with the transcendence of biological necessity and so with the emergence of values, politics, and the good. Yet materialism understood in this minimalist sense bears no meaningful resemblance to the US polity before the 1960s nor to a variety of cultural contexts in the Global South today—locations where Inglehart and Nordhaus and Shellenberger have suggested that it has been dominant.

In contexts like these, "materialism" must mean something else to these authors. Conceived in this second manner, it is a particular *kind* of politics that values and prioritizes questions of economic production and distribution, material livelihood, and connection to the everyday issues that emerge, as environmental justice activists have described it, "where we live, work, and play."[48] But if this is what is meant by materialism, then "postmaterialism" would have to entail a rejection of these preoccupations. It could not do so by proclaiming itself uniquely political or moral but only by opposing its politics and morals to the quotidian ones.

Understood in this way, of course, postmaterialist environmentalists can easily appear elitist and disconnected from the concerns that animate the less affluent—precisely the appearance that Nordhaus and Shellenberger proclaim that they reject.

The bottom line is that these two views of materialism—first as bare life and second as a preoccupation with livelihood and everyday life—generate mutually exclusive accounts of the dichotomy between materialism and postmaterialism. Yet both are partial and problematic. Neither by itself allows us to make sense of contemporary environmentalism, which might explain why Nordhaus and Shellenberger slide back and forth between them at different points in their analysis.

Materiality is central to understanding the dilemmas of environmental politics, but conceiving these in terms of the post/materialism binary—whether explicitly or implicitly—led this conversation down a dead end. If we accept the dichotomy between materialism and postmaterialism, then the sort of slippage we find in *Break Through* will be inescapable. We must instead recognize that everyday life is the location for concern for material needs, but these are always fulfilled (or not) in a value-rich context.

We must, in other words, reimagine *materiality* in a more inclusive manner, in which the biological and the cultural, problems and values, the concrete and the abstract are not opposing poles but necessarily entwined and inviting our attention. This would entail a rejection of the post/materialism binary, whereas Nordhaus and Shellenberger's elisions and contradictions result from trying to reverse what they see as the dominant valuation of one side over the other.

Reconstructing a Politics of Materiality

We must learn to talk about supposedly "objective problems" and "subjective values" together and to do so in a way that denigrates neither the reality of the problems nor of the values. In this way, we might encourage the development of a movement that is deeply committed to salient concerns of everyday life while recognizing the inherently value-laden character of these concerns. The role of experience is vital, but its meaning is never transparent. Whether we are discussing a toxic river fire, a drought, a forest clear-cut, or a new oil pipeline, the point cannot be—as postmaterialism would have it—to dismiss objective causes and consequences but to recognize them as subject to moral and political debate. Conversely, we must recognize the material reality of such events and issues without imagining that this reality can trump such debate. In this sense, the challenge is not to separate objective problems from subjective values—a separation that can never be achieved—but to ensure that material concerns do not enter the political realm in the wrong spirit.[49]

What does this wrong spirit look like? It is an instrumentalist approach that seeks to evade the turmoil of political contest by asserting the objective and unmediated authority from some other realm or source—physiological need, definitive science, absolute reason, or nature-imposed limits—to trump all else and insist that what is defined as an "objective problem" will therefore have an objective, singular solution. This is a tendency that is frequently manifest in environmental politics, though at its core it is antipolitical. It is an attempt to deny the inescapable role of politics as a means of negotiating difference.

The alternative is to enter political dialogue to advance a social vision that engages the values and concerns of people's everyday experience while also recognizing the ways in which the understanding of those concerns are enmeshed in a broader matrix of fears and insecurities, hopes and possibilities. Such vision must be also deeply grounded in the material conditions, because these are revealed to us, in part, by manifestations of need and by science. Problems and values are no longer imagined as competing bases for action but necessarily come together here to be better understood, discussed, and debated. This emerges from an understanding of the empirical and conceptual limitations of the approaches discussed throughout this chapter.

Yet at the same time there is also a normative case to be found here. That case centers upon the challenges regarding environmentalist identity and the definition of concern, which I outlined in the beginning of this chapter. An environmental politics focused upon materiality as subject to political contest ought thereby to accept plural foci, divergent interpretation, disagreement, and debate. It ought to draw the broadest set of actors and identities into dialogue, thereby inviting an expansive conception of what counts as an environmental concern.

One promising—though not unproblematic—resource for thinking through this contested politics of materiality is the rapidly growing body of work that has come to be labeled the "new materialism." Diverse in its approaches and disciplinary registers, this work is distinguished not simply by the primacy with which it regards material conditions but by its conceptualization of matter as interconnected assemblages of human and nonhuman, living and nonliving, and hence the self as transcorporeal—never existing in isolation from this broader network.[50] Agency, from this perspective, is not the product of a rational and sovereign self; indeed many argue that it is not a distinctive quality of humans—or even living beings—at all.[51] Agency is instead conceived on a continuum, with a wide variety of "actants" manifesting it in various ways and to varying degrees.[52] New materialism thus positions itself as a reaction to an excessive preoccupation with text, language, and social construction said to characterize other recent theoretical work while also promoting a broader conception of matter than the productivist and economic focus of Marxist materialism.[53] This attention to interconnected manifestations of matter is promising because it explicitly recognizes our embeddedness in the material world. Moreover, the emphasis upon diverse sources of agency calls into question the divide between subject and object. For these reasons, new materialism has the potential to cultivate more self-consciously political forms of action that invite material concerns into the public sphere in the "right spirit." Whether it fulfills this potential is the subject of the present section.

Vibrant Matter and Ontological Transformation

Despite the promise, one striking characteristic found in much new materialist writing is that it focuses upon urging a change in our worldview or paradigm—what many term ontology—to the near exclusion of the sort of pluralistic political contest sketched previously. Understanding both the attraction of this ontological focus and the ways in which it crowds out politics is therefore necessary. To that end, I begin with a consideration of political theorist Jane Bennett's book *Vibrant Matter: A Political Ecology of Things*, which offers one of the more engaging and widely discussed presentations of a new materialist approach. Bennett cultivates our appreciation for what she terms the vibrancy or vitality of matter in contrast to the fixed, mechanical, passive qualities often attributed to it. She argues persuasively that nonhumans and things "are vital players in the world" and offers a variety of stories intended to edify the reader about our necessary embeddedness and participation in this world with "the hope . . . that the story will enhance receptivity to the impersonal life that surrounds and infuses us, will generate a more subtle awareness of the complicated web of dissonant connections between bodies, and will enable wiser interventions into that ecology."[54] Bennett invites the reader to see the world, and our place within it, anew and to cultivate a new ethos rooted in this sense of belonging and connectedness. In these respects, her work provides an appealing set of reflections upon the material world in all its complexity and an appropriate introduction to new materialist writings. Nonetheless, I focus upon more troublesome characteristics of her argument in an effort to think through the relation of materiality to everyday life.

One striking characteristic of her book is how much of it is devoted to the development of what she terms "onto-stories," which might help us to "picture an ontological field without any unequivocal demarcations between human, animal, vegetable, or mineral. All forces and flows (materialities) are or can become lively."[55] By contrast, very little of the book (roughly eight pages) is devoted to addressing the question of "the implications of a (meta)physics of vibrant materiality for political theory."[56] My intent, in drawing attention to this point, is not to criticize Bennett for the political theory that she doesn't develop but to better understand why even a book-length account by a political theorist would be so disproportionately focused upon ontology.

The answer, for Bennett, is that a change in our ontology is both a necessary and difficult precursor to any reconstructed politics—and, in particular, to a reconstructed environmental politics. In this sense, the attention to ontology is misunderstood if it is read in the traditional philosophical sense as a metaphysical quest to comprehend properly the nature of being. As I noted at the outset of this section, ontology here is understood as synonymous with the way others describe a worldview or "paradigm." It serves to delineate the boundaries and character of comprehension and common sense among a broader population.[57] Ontology in this sense draws our attention to the ways in which people make sense of their world and their relationship with others, thus bringing it surprisingly close to the social-psychological conceptions that scholars like Inglehart seek to characterize.

The "onto-stories" Bennett tells—about encounters with the things she finds in a storm drain, an electrical power grid in collapse, the material flows of food, and others—are aimed to affect readers in ways that will cultivate a new ontology based upon new ways of perceiving and experiencing the material world. And yet, she emphasizes time and again, this is hard:

> Even if, as I believe, the vitality of matter is real, it will be hard to discern it, and, once discerned, hard to keep focused on. It is too close and too fugitive, as much wind as thing, impetus as entity, a movement always on the way to becoming otherwise, an effluence that is vital and engaged in trajectories but not necessarily intentions. What is more, my attention will regularly be drawn away from it by deep cultural attachments to the ideas that matter is inanimate and that real agency belongs only to humans or to God, and by the need for an action-oriented perception that must overlook much of the swirling vitality of the world. In composing and recomposing the sentences of this book—especially in trying to choose the appropriate verbs, I have come to see how radical a project it is to think vital materiality. It seems necessary and impossible to rewrite the default grammar of agency, a grammar that assigns activity to people and passivity to things.[58]

A radically difficult project for the author—perhaps an impossible one, she suggests in this passage, for society at large. Yet Bennett hopes that "a discursive shift from environmentalism to vital materialism [will] enhance the prospects for a more sustainability-oriented public."[59] The focus on ontology becomes comprehensible when her challenge and her aspiration are framed in this manner. Such a "necessary and impossible" task cannot be dispensed with quickly or lightly. Nor can it readily be entrusted to others—a conclusion that becomes clearer once we appreciate just what sort of project Bennett's book exemplifies.

Defining the effort as one that requires us to "*think* vital materiality" anew makes it plain that ultimately this is less a materialist project than an idealist one.[60] As a consequence, it returns us to some of the dilemmas of the "postmaterialist" account of environmentalism in which—as in Bennett's account—the focus was upon the difference that thinking anew makes, upon the character and transformation of the "subjective values" that we use to make sense of our world.[61]

The limitation of an ontological approach to vibrant matter might be further illuminated by drawing out the understanding of the theory–practice relationship that it appears to embody. Bennett seeks to convey just how hard it is to think of matter anew. This emphasis upon difficulty is a consistent theme in the book, which comes to a head in her conclusion. On the last page, Bennett presents herself as exhausted from the task, proclaiming that although some of her arguments are "pitched at a very high level of abstraction or generality" and although "more needs to be said," "I am, for now, at the end of my rope."[62] She ends the book with what she describes as "a litany, a kind of Nicene Creed for would-be vital materialists":

'I believe in one matter-energy, the maker of things seen and unseen. I believe that this pluriverse is traversed by heterogeneities that are continually *doing things*. I believe it is wrong to deny vitality to nonhuman bodies, forces, and forms . . . I believe that encounters with lively matter can chasten my fantasies of human mastery, highlight the common materiality of all that is, expose a wider distribution of agency, and reshape the self and its interests.'[63]

If the theorist's task is hard, the challenge of putting it into practice is even harder. Yet the analogy to the Nicene Creed is revealing. The sequence of events she envisions now seems clear: first *we* theorize the vibrant matter "creed," then we spread the gospel so that *others* can act upon this good news. Practice follows from theory—and theorists. We must get first things right first. This approach epitomizes that of the "outside critic."

Engaging with matter can be edifying and illuminating, but if the goal is "a more sustainability-oriented public" and more efficacious ways of discussing environmental challenges, then it need not—and ought not—focus upon the task set out by Bennett. It need not because, properly understood, "encountering a vital materiality" is not something new to any of us. It ought not, because a work that sets out to transform something characterized as a social ontology is more likely to be received by other members of the society as patronizing and paternalistic than enlightening or consciousness raising. Our challenge must be to work with popular values and concerns to cultivate a meaningful politics of sustainability and resilience. That requires theory to begin by engaging materiality as it is already manifest in practice. This can cultivate openings for challenges to such practices from within rather than attempting to transcend or transform those values from without.

Materiality of Everyday Practice: Political Not Metaphysical

I now turn to a different manifestation of the new materialist approach. The difference may seem subtle, and it is not explicitly distinguished by the authors involved. Yet it is a difference that *makes* a difference—especially if one is motivated by a commitment to advancing resonant social criticism. Here, instead of characterizing the recognition of materiality as the result of a very difficult ontological transformation it emerges from reflection upon the everyday. It is, in the words of Stacy Alaimo, "never an elsewhere but is always already here."[64] The consequence is that it is more likely to resonate with the experience of those whose practices are described and less likely to be perceived as the moralistic prescription of an outsider.

One compelling approach to this sort of everyday materialism is to focus upon the manifold ways that "humans are the very stuff of the material, emergent world."[65] Alaimo explores the reflections, memoirs, and activism of those whose experience of their own bodies are entwined with the often toxic consequences of the broader material world (such as those with multiple chemical sensitivities). Here it becomes clear that environmental health, justice, and toxicity are more than discourses; these products of everyday life and struggle are literally embodied. By highlighting such embodied knowledge, she recognizes and draws out the insights to be had from people who are infrequently heard or taken seriously in scholarly discourse.[66] This approach tempers what Teena Gabrielson and Katelyn Parady describe as the frequent and troubling tendency in environmental thinking to privilege "those positioned *to know or imagine* a particular conception of what a green 'good life' would entail," at the expense of more experiential and embodied forms of knowledge—a privileging, I have suggested, that is integral to the call for ontological transformation.[67]

The materiality of everyday practice, however, is not only manifest in our bodily nature. In her essay entitled "Plastic Materialities," Gay Hawkins—like Bennett—explores the material resonances of an everyday thing—in this case single-use, "disposable" plastic shopping bags. In discussing these bags, Hawkins, too, tells stories; in fact she tells three distinct and potentially contradictory ones. Unlike Bennett's stories, however, Hawkins's are not presented as difficult to apprehend and if anything seem distinguished by their familiarity and even their utter banality.

The stories highlight ways in which plastic bags are—or have become—"contested matter."[68] In the first account, Hawkins sketches environmental efforts to reduce or eliminate plastic bag use in stores; what she describes as "Say No" campaigns alters—arguably in small ways—the everyday practice of shopping and transporting food and other consumer goods. Here "specific aspects of the materiality of the plastic bag" are highlighted: "their slow process of decomposition, their tendency to trap or choke marine animals, their oppressive ubiquity."[69] This list could easily be lengthened, and given their ubiquity—an estimated one trillion bags per year—the magnitude of the harm of these plastics to ocean life is tremendous. Precisely because of the damage caused by these material effects, Hawkins rightly observes that such campaigns entwine the everyday experience of shopping with "circuits of guilt, self-reproach, and virtue."[70] This is an important point to which I will return.

The second account is of the sort perhaps even more familiar, at least to those immersed in the practices of middle-class parenting in postindustrial societies. Here Hawkins presents the bag as a quick solution to the problem of transporting a child's wet swimsuit, as well as resolving a parent's last-minute anxiety as the family hurries out of the house on a school day. The bag's ready availability, its easy portability, its waterproof plasticity, its "humble practicality" manifests other material qualities that generate a different affective response among those responsible for such household practices: relief, even gratitude.[71] Of course, given the frequently gendered division of such household responsibilities, the parent feeling such relief is far more likely to be a woman.

The third moves from everyday practices to a widely-viewed Hollywood movie, *American Beauty*, discussing a video portrayal within the movie of a plastic bag blowing in the wind, generating—for the characters in the movie and perhaps for the viewer—a response to the bag as an object of surprising, near-transcendent beauty. Despite—or perhaps because of—the cinematic choreography, material qualities of the bag are again central: its light weight, ability to catch and hold the wind, the viewer's recognition of the ubiquity of such bags and so of the likelihood of finding one lying about—or floating above—a city street.

Hawkins uses the fact that such mundane objects often generate diverse and potentially contradictory framings, responses, and affects to draw attention to the problem that she—following Noel Castree—describes as "material essentialism."[72] This is the presumption that things have a singular and relatively immutable set of properties and associations that define them. From the perspective of the "Say No" campaign, for instance, the sense is that the environmentally destructive qualities of plastic bags are the *only* relevant properties of bags, which is also manifest in the frequently moralistic tone of the campaigns. Hawkins's approach, by contrast, demands that we recognize the complexity and plurality of material relations, even with a thing as seemingly uncomplicated as the bag. After all, if one is seeking to ban plastic bags due to their ecological impact, a failure to recognize and so grapple with their practicality and convenience is to disregard some of their more politically salient affects. These affects are differentially distributed throughout society, because a classed and gendered division of labor means that changes to a practice such as plastic bag use will have greater impact on those—predominantly women—who are much more likely to bear primary responsibility for food shopping and quotidian parenting tasks, such as helping a child with a wet swimsuit get ready to depart for school. This can shift the discourse about banning plastic bags away from an all-too-easy rejection of wasteful consumerism and toward meaningful engagement with the roles and responsibilities associated with these material practices.[73] Rather than following Bennett in seeking a new worldview based upon an unfamiliar new way of perceiving matter, Hawkins calls to our attention already familiar and diverse ways of relating to matter and thus the diversity of ways in which a call for change is likely to be received. Only clear-eyed attention to this can position us to draw out the normative implications and challenges of everyday practice.

There are two important consequences of this approach. First, by fostering a recognition that we always already engage with the material world it becomes more viable to argue that those forms of engagement need to be addressed. Second, by illuminating the inescapable plurality of our experience and complexity of our values it subverts the temptation to moralism that has often generated a backlash of resentment in environmental campaigns.[74] Plurality, moreover, allows us to imagine alternatives to present practices that also draw upon threads of everyday experience. This creates space for a critical theoretical approach to emerge immanently from within our practices.

It is crucial to be clear here. By itself, Hawkins's recognition of the plurality of material effects and affects of plastic bags offers no guarantee that it will advance a critical perspective. Indeed, some will worry that it would result in a paralytic leveling or an excuse that allows one to believe that everyday convenience or the rare opportunity for beauty can somehow counterbalance the petrochemical composition of bags or their harmful effects on marine life. Yet beginning with a commitment to social criticism this recognition will be key to shaping the critical agenda in a manner immanent to the material practices under consideration. The harms caused by our ubiquitous use and disposal of the bags are not a social construction and are not somehow tempered by these other effects. Plurality does not justify inaction but does promote recognition of the complex attachments many of us have to things deemed by some—even by us!—"bad."

Without such recognition, diagnosing the obstacles to social change is far more difficult. Moreover, resentment and resistance to moralizing become prominent obstacles. My aim is not to deny the importance of morality or—more broadly—the importance of normative argument. It is instead to indicate how and why the temptation to sanctimony and moralism impedes normative argument from resonating with listeners. Morality without moralism is both viable and necessary. Normative argument that recognizes complexity and plurality is less likely to foster backlash. Rather than imagining that "the" right argument can win the day, we must develop arguments rooted in an ethos of real respect for the everyday experiences of those with whom we wish to engage in dialogue.

Contested Materiality Creates Opportunities for a Democratic Public

It is necessary, then, to escape the binary between the objective and the subjective, between materiality and values. With Bruno Latour, we must seek "never to get *away* from facts but *closer* to them"[75] yet to recognize that the closer we get to what is real, the less fixed and certain and objective it appears. Latour makes this point in a self-critical essay that challenges some of the tropes in his own work, which has often been read as presenting what scientists and others took to be reality as nothing more than a social construction. Latour now worries that an undue emphasis upon the human and constructed qualities of our accounts of the world misleadingly invites the conclusion that all accounts are merely subjective. He argues convincingly that nothing could have been further from his intention. Yet he argues emphatically that this notion of arbitrary subjectivity cannot truly be rejected by replacing it with a naïve or rigid assertion of objectivity. This allows Latour to simultaneously affirm a commitment to facts and to argue against what he calls "matters of fact." His argument against the latter is that to regard the meaning of material things as transparent, determined, or given—to view them in the way Inglehart characterizes "objective problems"—is not a realist position at all, but instead these matters of fact are "very partial and, I would argue, very polemical, very political renderings" of what are more accurately characterized as "matters of concern."[76] Matters of concern, in this sense, are the contested materiality I have sketched here. They are a path to respectful recognition of the materiality of things, which requires that we not misrepresent their import by presuming that their meaning is fixed, singular, and, thus, apolitical.

In sum, to reimagine the politics of environmental sustainability as a contested materiality that seriously engages the complex web of material practice we must navigate between the Scylla of the postmaterialist interpretation in all its forms and the Charybdis of a material essentialism. To do so doesn't only displace the postmaterialist account of the origins and growth of environmentalism. It also creates possibilities for citizen action inclusive of a more diverse set of identities, experiences, and embodied realities and thus for concerns defined in a far more pluralistic manner.

Yet it remains a bit cavalier to describe this as "politics" at this point. The dilemma is that once we take seriously the idea of engaging materiality we are compelled to address practices that are today closely associated with notions of privacy and a private sphere: activities within the home, conceptions of private property, and the world-transforming impact of private automobiles to name just ones that will be the focus of attention later in this book. Before we focus on these practices, however, it is necessary to challenge received accounts of the relation between public and private so that these too don't impede thinking or acting politically upon "private" issues. Given the project I've sketched in this chapter and the last, navigating public and private will require escaping the strictures of the discourse of liberalism while formulating a way to address such material practices as fundamentally public concerns and simultaneously respecting the private character of human experience in these aspects of everyday life. It is that task that I take up in the next chapter.

4

Private Practices yet Political

The contested politics of materiality outlined in the previous chapter can quickly run up against a conceptual wall: in a variety of ways, material practices—including those that have the biggest ecological impact—are typically construed as *private* in western societies, and as such neither public nor political.

Consider some illustrations related to cases that I examine in later chapters. "A man's home is his castle" is a familiar idiom derived from English common law (ostensibly universalized from "an *Englishman's* home . . ."). It has deep roots there and in the US legal tradition (reflected in the Fourth Amendment to the US Constitution) and reflects a sense of home as a private sphere protected from public intervention.[1] Private ownership of land, likewise, is a quintessential example of something construed as protected from the intervention of the state and other individuals. Automobiles provide many with an encapsulated space of exclusion and privacy, unlike both "public" transportation in the form of buses and trains and even walking on "public" sidewalks. Consumption choices—about food, energy, and consumer products—are also paradigmatically regarded as private ones, both in the sense that they are viewed as economic choices in the "private sector" and in the sense that they reflect choices rooted in personal identity, belief, or preference.

This widespread sense of material practices as not just private but archetypically so presents two related challenges. First, the consequence is that they also do not seem to be the stuff of politics. This, of course, is not an empirical truth: autos, land, homes, and consumer goods and transactions are subject to law and regulation in many ways. The point is that there is often a default bias that must be overcome by those promoting such law or action because of the degree of invisibility or counterintuitiveness to the role that they play in constituting these "private" practices. If we are to talk about a politics of material practices, we must first unmask this invisibility. Second, as the preceding examples are likely to remind us, the private is often associated—deeply yet problematically—with notions of individual freedom. This becomes a normative basis for resistance to such a politics, opposing the violation of this freedom.

The challenge, then, is to respect—and at times even to expand upon—the sense of privacy and freedom that is so often associated with everyday material practices while simultaneously opening them up to more explicit political contestation. To do so requires that we recognize the diverse forms of exclusion from these opportunities for privacy and freedom that are the everyday experience of many and recognize that these exclusions are no accident but instead reflect the exercise of power. We must also highlight the often unacknowledged constraints that even those of us who are privileged enough to benefit from these practices (driving, living in comfortable homes, owning property, etc.) regularly experience.

In recent decades, many feminist theorists have articulated a powerful critique of the exclusions and forms of privilege that are bolstered by depoliticized western conceptions of the private sphere, particularly as this relates to gender roles and the home. A variety of other critical theorists have also challenged the boundaries drawn between public and private, highlighting the ways this shields economic activity and technological development from political engagement. My argument is heavily influenced by these analyses. In this chapter, I also draw upon two figures in the tradition of philosophical pragmatism—John Dewey and Karl Polanyi—in order to highlight the ways in which everyday experiences in private life can and often do prompt the formation of a public (or publics). In this way, they position us to see beyond the dichotomy of public and private *spheres*, to recognize the always already political character of everyday material practices, while also recognizing the appeal of privacy and freedom as motivations for public action. This positioning is invaluable to the task of engaged, insider critics, because it enables a critique of current relations of power and privilege that build upon positive associations of material practices with the private.

For example, the saying that a man's home is his castle does not simply reflect a belief about home as a space of privacy and freedom; it simultaneously ought to draw our attention to the gendered exercise of power relations at work within this private space and protected by this notion of privacy and freedom.[2] In many ways, the freedom of women and of children is often compromised by this exercise of power in the home. Similarly, the identification of land ownership with freedom is necessarily partial and requires recognizing that such ownership has only been available to a minority, thus simultaneously manifesting this freedom as a form of exclusion and inequality. Private automobiles propel their drivers and passengers along "public" streets, access to which is restricted for those not so encapsulated.[3] Thus the freedom enabled comes at considerable disadvantage to those without access to cars or unable to drive, as well as to all of us when we seek to navigate through space by using alternate forms of mobility, including bicycles, buses, or on foot. Finally, with regard to consumer products choice is premised upon time, as well as market access, which is of course dependent upon highly—and increasingly—unequal control of wealth and income. A critic who highlights these limitations on freedom, privacy, and choice can argue for their more complete realization rather than rejecting these values altogether.

To foster a politics of material practices requires a more encompassing notion of the public. The prospects for this become greater when we recognize that it need not come at the expense of freedom in private life but can often facilitate more meaningful and diverse forms of freedom that are inclusive of more members of society.

The Centrality of a Public/Private Dichotomy

Drawing the lines between public and private—both practically and theoretically—has been a central preoccupation of Western thought since classical antiquity.
—Jeffrey Weintraub[4]

The profound ambiguity of the liberal conception of the private and public obscures and mystifies the social reality it helps constitute.
—Carole Pateman[5]

The dichotomy is not one, but deploying it as one has material and cultural effects.
—V. Spike Peterson[6]

Conceiving material practices in western societies as private ones is enmeshed with a familiar dichotomy between private and public. The result is that material practices are regarded as located in a discrete and bounded *sphere* and that they have a *character* that is regarded as unambiguous and uncontested. A closer look, however, reveals that although a public/private dichotomy is ubiquitous in everyday conversation and in academic discourse, its character is equivocal and shifting. Certainly, it is a dichotomy that permeates liberal theory (though, importantly, not only this as I go on to discuss), and so I turn first to its manifestations in this context.

Liberal political theory that traces its roots to John Locke typically privileges the private as a sphere that exists outside of, and prior to, the formation of the public. In this Lockean sense, nature is conceived as prepolitical, and mixing one's labor with it becomes the basis for justifying private property as a natural right.[7] Nature, labor, and property are at the heart of what I have described as material practices. Not all liberal theorizing follows Locke in conceiving the private sphere as ontologically prior. However, most such theorizing seeks to defend a private sphere from outside intervention—from the state, to be sure, but also, as Raymond Geuss notes, "from all kinds of intrusion . . . 'Private' here means . . . 'what *ought not* (for whatever reason) to be interfered with either by other individuals or by social and political institutions or agencies.'"[8] In liberal theory, that which ought not be interfered with includes attitudes and beliefs, especially as they relate to conceptions of the good life.[9] But protection of these has been intimately entwined with "private property"—a material sphere of control, exclusion, and hence privilege for its possessors.[10] The identification of noninterference at the conceptual heart of the private sphere is also characterized as a negative conception of freedom.[11] Freedom, here, is construed as the ability to exercise our "private" choices, in the sense that they are made "free from" interference by the "public."

This familiar association of private choices—and a material, private sphere—with freedom from interference is deeply problematic. This conception of the private sphere effectively shields the exercise of power and privilege there from public scrutiny or sanction. Marx and Marxists have long emphasized its centrality in the realm of social and economic activity.[12] Feminist theorists have greatly advanced our understanding of this gendered exercise of power within the household and family.[13] Undertheorized by both, however, are the complex ways in which other material practices and assemblages also shape and define the limits and possibilities of freedom. The liberal association of the private sphere with noninterference remains strong, however, and is often the basis for conceiving public or political action as a threat to the private and hence a threat to freedom. By first recognizing the role this association plays within liberal theory and then the substantial limitations and exclusions from these exercises of freedom, we will be better able to identify constituencies, contexts, and resources that can be a basis for effectively challenging this association.

A public/private dichotomy is not unique to liberalism and is manifest in other theoretical traditions as well. For example, Hannah Arendt seeks to reconstruct and revive a preliberal, civic republican sense of this dichotomy, which she does by appealing to the ancient Greeks and to Aristotle in particular. Arendt—arguably the twentieth century's preeminent theorist of public and private—frames the divide as between *oikos*, the private realm of the household economy and family, and *polis*, or the public, common world of political community.[14] The former, in Arendt's construction, was the realm "born of necessity," and mired in inequality and violence. The latter was the realm of equality and hence freedom—understood first and foremost as freedom from being ruled.[15] Here, the public sphere is synonymous with politics, though its association with freedom from being ruled makes it clear that politics cannot be readily identified with government or administration. As with liberal theory, Arendt's account of the dichotomy seemingly places material practices decisively in a sphere of private life while reserving politics and citizenship for the public sphere. Although Arendt recognizes that these private practices make the public possible, this public is characterized as wholly separate from them.[16]

Arendt does not, then, diverge from liberal theory in her account of the centrality of this public/private dichotomy, nor in the essential content of each side of the divide. Where she diverges is in the normative revaluation of liberal theory. She valorizes the public as the only sphere in which meaningful freedom is possible. By contrast, she offers an account that devalues practices in the private sphere—the sphere of privation—in which the exercise of such freedom appears nonsensical.[17] Yet Arendt does not end her analysis with this binary. Instead, she offers a historical account in which it has evolved into a more complex, tripartite division, arguing that in modernity "the social realm" has arisen by drawing in to it much of the earlier content of both the private and the public realms, an important claim to which will I return.[18]

For now, despite many other differences, we can observe that the conception of material practices that emerges through these otherwise divergent accounts of public and private has two central features. First, they are characterized as existing—spatially and conceptually—in a *sphere* or *realm* that is private—that is, a physical space separated from "the public" and the location where privacy is experienced. Private and public are thus spatially separated and mutually exclusive. Second, the *character* of material practices in this private realm is described in a distinctive manner. Practices are not subject to evaluative judgments that appeal to shared criteria, nor are they subject to debate, negotiation, or other forms of contestation—all manifestations of politics, broadly construed. Instead, activity in this private realm is characterized as unequivocal, determined, objective, or natural. In this case, it is consistent with the exercise of individual interests and might be deemed appropriate for purely technical adjustments by experts, but it is inconsistent with the character of political decision making by citizens.

By conceiving both the space and character of material practices to be private in these senses, a particular range of alternatives comes to be seen as feasible and appropriate, whereas other alternatives are not. Where material practices are conceived as constitutive of a private sphere exclusive of, and privileged over, the public, problems of sustainability appear to require some sort of violation of, or intrusion into, this sphere. To the extent that the character of private practices is viewed as fixed by the objective facts of the matter, however, appropriate change would appear to preclude a self-consciously political response. Instead, it seems to demand an outsider who can either initiate an expert-led technocratic intervention or some sort of cultural transformation leading to the adoption of a radically new identity, resulting in new attitudes or preferences.

Politicizing Material Practices

We must overcome the limitations of this confined and undemocratic way of thinking of alternatives, one that makes environmental *politics* seem like an oxymoron. To do so requires critiquing this view of the private. The challenge is to see if and how material practices associated with private life can be appropriately construed as political while also respecting the continued appeal of freedom and privacy. We can begin by disaggregating the notion of a private *sphere* from notions of the *character* of private practices. A brief return to Hannah Arendt is illuminating here.

As noted previously, Arendt's initial account of a dichotomy between public and private spheres is made more complex by her account of the rise of an overlapping third sphere that she terms "the social." In her analysis, "the social" threatens to engulf both the private and public in contemporary society. *Society* emerges when "housekeeping, its activities, problems, and organizational devices" rise out from the private household.[19] Central to these activities, problems, and devices are those of what we have come to know as the economy. What is distinctive and relevant about the modern economy in contrast to the ancient *oikos*, in Arendt's view, is that "behavior" rather than action becomes the norm, resulting in regularities that can be measured and predicted with statistical confidence.[20] The rise of the social thus describes both a loss to the private sphere and a displacement of the public sphere in which political action is said to be central. The result here is a replacement of political life with "pure administration."[21]

It is clear that this social realm is, for Arendt, where we would now expect to locate many material practices—largely displaced from what she views as the private per se, though tellingly often labeled as the "private sector" in western societies. It is equally clear that the character of this rapidly growing realm—regularized, normalized, measurable, and hence amenable to technocratic administration—threatens to engulf the earlier political notion of the public completely.[22] The problem of the contemporary age, in Arendt's terms, is the problem of the uncontrolled growth of the social—where this is understood both as a *realm* of material practices and by its routinized and pliable *character*—subject to measurement and effective control by outside experts.

Consistent with Arendt's account, many theorists have observed that it is in such a technocratic realm that instrumentalist approaches have dominated the criteria for decision making on environmental challenges.[23] Yet as Ulrich Beck and Tim Luke have shown, technocracy is unable to contain a growing struggle over the decision-making criteria, leading to the eruption of what they term "subpolitics"—a surreptitious manifestation of the contested politics of materiality that I named earlier.[24] The actual character of the struggle that emerges here is thus hard either to recognize or acknowledge given the familiar associations of politics with an institutionalized, publicly identified space.

At this point, Arendt's account seems simultaneously familiar and troubling: familiar in the sense that it too conceives of material practices as existing as a hybrid, neither evidently public nor private but something in between,—troubling in the sense that "the social" demands our vigilance and concerted effort to hold it back from the continual danger it poses to the vital but unappreciated role of the public in the creation of a common life for the community. If this is right, then the contested politics of materiality I have prescribed—and that Beck and Luke have described—becomes a manifestation of the problem. After all, such a "politics" would become a portal through which the material realm infiltrates and contaminates the purity of political action.

But is this really what Arendt is trying to tell us? One of the most perceptive interpreters of Arendt, Hanna Pitkin, argues that it cannot be. If the public were to be successfully shielded from all manifestations of the social, Pitkin points out, then political action would cease to be about *anything* meaningful to our lives. It would be little more than a venue for "posturing little boys clamoring for attention ('Look at me! I'm the greatest!' 'No, look at me!') and wanting to be reassured that they are brave, valuable, even real."[25] Although we may find that this image cuts too close to actually existing political activity, it hardly seems possible that Arendt would wish to encourage or protect such expressions of vanity and anxiety. Moreover, in other places, Arendt points to the constructed, material world as that which "gathers us together" in a public realm: "a world of things is between those who have it in common, as a table is located between those who sit around it."[26] It is this world, she argues, that generates "specific, objective, worldly interests."[27] Pitkin argues persuasively that Arendt could not have intended to protect politics from the concerns and activities in the realm of the social but only from the usual *character* that she attributes to these activities—a character also reflected in her account of worldly interests as "objective." Inspired by Arendt, Pitkin concludes that "far from excluding the social question as unworthy of political life, we need to make it political in order to render it amenable to human action and direction. The danger to public life comes not from letting the social question in, but from failing to transform it in political activity, letting it enter in the wrong 'spirit.'"[28] Recognizing materiality as integral to what Arendt terms "the social question," Pitkin's reading of Arendt bolsters the case for the contested politics of materiality that I have sketched, emphasizing both the need to bring such concerns into political activity and the need to change their character in the process of doing so.[29]

Pragmatism, Practice, Privacy

The approach to material practices I am pursuing begins with familiar associations between materiality and private life yet simultaneously recognizes and embraces the ways that these practices are defined publically and politically. This can be appropriately described as pragmatic. It echoes key elements in the tradition of philosophical pragmatism in two ways: its grounding in *experience* and its attention to *consequences*.

First, for the pragmatist, respect for lived experience calls us to recognize the inseparability of the subject *having an experience* and the material things *being experienced*. John Dewey provides insight here. In a passage already quoted in this book's introduction, he argues that the true test of a philosophy's value is as follows:

Does it end in conclusions which, when they are referred back to ordinary life-experiences and their predicaments, render them more significant, more luminous to us, and make our dealings with them more fruitful? Or does it terminate in rendering the things of ordinary experience more opaque than they were before, and in depriving them of having in 'reality' even the significance they had previously seemed to have? Does it yield the enrichment and increase of power of ordinary things which the results of physical science afford when applied in every-day affairs? Or does it become a mystery that these ordinary things should be what they are; and are philosophic concepts left to dwell in separation in some technical realm of their own? [30]

A philosophy that renders experience "opaque," that exists only "in some technical realm" is what Dewey and other pragmatists rail against. They emphatically do not disdain theory or philosophy itself but reconceptualize its role by insisting that it must both begin and end with experience. Its value can be measured by its ability to illuminate this experience.

In the face of our present challenge—overcoming the constraints upon a politics of materiality imposed when the latter is conceived as private—one strategy is to simply reinscribe material practices as public instead. That is, we might leave unchallenged the spatial and conceptual duality between private and public while proposing to move private things to the public sphere. For example, as I discuss in the next chapter, some who identify obstacles posed by the private ownership of land propose that this simply be made public instead. The pragmatist recognizes the obstacles to such an imperial gesture. It fails to respect or understand the lived experience of citizens, in which the private frequently plays a powerful and attractive role—both as a practice and an aspiration.[31] In Dewey's terms, this move would deprive experience and ordinary things of their reality in favor of some ostensibly transcendent conception to be imposed from outside.[32] As such it is disconnected from the role that a theorist or social critic either can play or should aspire to. Although any particular boundary drawing between public and private must be regarded as a tentative settlement and a reflection of relations of power, it is also the case that claims to privacy are frequently seen by those who are disenfranchised or relatively powerless "as autonomy and the ability to set the terms of one's life course, [and] as a necessary condition of citizenship."[33]

In his premier work of political theory, *The Public and Its Problems*, Dewey argues emphatically against trying to comprehend the state or the public by looking to its causes—whether in human nature, divine creation, the social contract, the advance of reason, or elsewhere.[34] In this sense, although he is often labeled a liberal, his thinking takes us beyond the confines of this tradition as it is usually construed. Dewey is focused squarely on the consequences of our actions as the basis for differentiating private from public. His philosophical naturalism and respect for natural science predisposed him to regard materiality with the utmost seriousness,[35] yet it is his attention to consequences that places material practices on center stage. By directing our attention to the consequences of human actions, Dewey argues that the very definition of "the public . . . consists of all those who are affected by the indirect consequences of transactions to such an extent that it is deemed necessary to have those consequences systematically cared for."[36] The public is not preestablished by fiat of the state and—importantly—is *not* a fixed sphere or realm of life. Instead, it comes into being and distinguishes itself from the private for Dewey precisely when and where people recognize and respond to the material effects, or indirect consequences. This character of a public, for Dewey, is such that participation by the affected is vital and so democracy both desirable and necessary.[37] Neither liberalism nor any other theoretical tradition can adequately circumscribe the concerns or actions of such a protean public.

If public and private are not preordained and separate spheres of life, if this powerful and familiar distinction is nonetheless equivocal and plural, then how shall we understand their relationship to materiality? It seems to me that the answer most consistent with Dewey's approach is that public actions are those that in a particular time and place are conceptualized as matters demanding some form of collective judgment and accountability.[38] A public may judge a given practice to require greater publicity—as, say, in the case of the actions of corporate polluters. Alternately, a public may judge a practice as requiring the recognition and equal treatment of private choices, as in contemporary arguments for gay marriage, or the respect for privacy itself, as in arguments against government electronic surveillance. Despite the differences in prescription, all represent the action of a public.

It is in this sense, then, that philosophical pragmatism might inspire the actions of an inside critic. A truly pragmatic approach to criticism wouldn't seek answers from Dewey (or James, Pierce, or Rorty) for environmental action. Rather than offering insight into *what to think*, my reading of Dewey suggests that he is most useful in answering the question of what to think *about*: the consequences of our lived experiences in the world.

Theory versus Practice

What is this "public" of which Dewey writes? And what do they do? More particularly, for our purposes, what does it look like for an environmental politics to emerge as a consequence of material experiences? On these questions, Dewey is more difficult to pin down. Here, I turn to one additional mid-twentieth-century scholar for insight: economic historian and social theorist Karl Polanyi. Although Polanyi is not usually described as a philosophical pragmatist, his argument and analysis is consistent with and also often fleshes out the more abstract arguments about experience and consequences advanced by Dewey and other pragmatists. Thus it should not be surprising that Dewey wrote with enormous enthusiasm about Polanyi's 1944 masterwork, *The Great Transformation: The Political and Economic Origins of Our Time*, describing it as "intensely exciting," "a wonder,"[39] and "the most enlightening account of the important historical events in the last century-and-a-half I had ever read."[40]

The "great transformation" about which Polanyi wrote was the ambitious nineteenth-century effort to institute a laissez-faire capitalist economy, stripped of government regulation, tariffs, and other restraints. Disastrous in its effects upon society, according to Polanyi, this effort was doomed to failure and in fact collapsed by the outbreak of the First World War. Yet Polanyi's account is not merely of historical significance. As interpreters have noted, neoliberal advocates of globalization in recent decades "embrace the same utopian vision" articulated by the leaders of Polanyi's great transformation;[41] indeed we appear to have undergone a "second great transformation."[42]

For present purposes, Polanyi's greatest insight is in his conceptualization of the "countermovement" that can emerge in response to this transformation. The countermovement manifests the characteristics of a Deweyan public better than anything Dewey himself offered, thereby fleshing out a crucial but notoriously ambiguous aspect of his political theory.[43] For Polanyi, the role of the countermovement can only be understood by following his emphatic distinction between the political-economic *theory* of liberalism (both past and present) that often drives both our understanding of existing conditions and the movement for economic and political liberalization versus the actual *practice* of nation-states and international organizations claiming to implement change based upon this theory. Whereas the former imagines an expansive private sphere but very narrow boundaries within which political action takes place, the latter reflects an expansive—albeit often unrecognized—political sphere that is essential to the constitution of any notion of the private.

To understand why, in Polanyi's account, we so often mistake theory for practice requires an understanding of his widely employed concept of "embeddedness." As interpreters have noted, at the core is "the defining dichotomy of the embedded economy and the disembedded, or autonomous, market."[44] Polanyi draws upon Aristotle's argument about the necessity of a conception of the *good*—and of the ends of a good life—for the structuring of livelihood and economic activity. Aristotle describes the process of economic acquisition as *instrumental* to a conception of the good and thus argues for the importance of *limits*: "there is no art that has an instrument that is without limit."[45] This enables him to distinguish the sort of acquisition properly pursued by household managers and political rulers—limited by the end or use to be made of it—from "expertise in business . . . [for] which there is held to be no limit to wealth and possessions."[46] For Polanyi, this Aristotelian distinction between "householding" and "moneymaking"—between acquisition for use and for gain—"was probably the most prophetic pointer ever made in the realm of the social sciences" and his own embedded–disembedded dichotomy clearly builds upon it.[47] Thus, an embedded economy is one modeled after the household, in which acquisition is a process of securing use-values, where the ends of the community itself define these values. To seek to disembed an economy is thus to model it after Aristotle's characterization of what is "held to be" business expertise, in which the absence of an understanding of the properly instrumental role of acquisition leads to the absence of "limits to wealth and possessions."

With this understanding of embeddedness, we can better appreciate its connection to the materiality of our lives and relations, a point Polanyi also emphasized. He develops this connection by first distinguishing between two meanings of the word "economic" that often have been intermingled in our contemporary uses of the term. These are the "formal" and the "substantive" definitions. The formal definition has its roots in "the logical character of the means-ends relationship, as in *economizing* or *economical*."[48] Here, we would expect to find *homo economicus*: the calculating human driven by scarcity to maximize his or her individual well-being. This definition regards all forms of acquisition as unlimited by use-value or a conception of the good and is, in Polanyi's view, a highly culture-bound and ultimately illusory understanding of human behavior. By contrast, the substantive meaning of economic is especially relevant to an appreciation of the role of materiality and everyday life. This meaning "points to the elemental fact that human beings, like all other living things, cannot exist for any length of time without a physical environment that sustains them."[49] This reflects the human condition itself: "no society can exist that does not possess some kind of substantive economy."[50] The confusion (or misrepresentation) of this with the formal conception is what Polanyi terms "the economistic fallacy."[51]

There is a normative point here, as well as the empirical one: if human behavior is frequently embedded within a web of material and social relationships that predispose us away from the narrowly self-interested calculations of *homo economicus*, then we can reject the claim that the latter is coextensive with human nature as such. Polanyi takes the point further to argue that the ideal of formally economic calculating individuals operating within a self-adjusting market is truly a "utopia"—it exists *nowhere*—and indeed "could not exist for any length of time without annihilating the human and natural substance of society."[52] To believe otherwise is, again, the "economistic fallacy."

In sum, although this notion of an autonomous private sphere of the fully "self-adjusting market" is prized and central to the *theory* of liberalism, it is vital to recognize that it cannot be fully realized over an extended time in *practice*. Thus embeddedness must always be a question of *what sort* of conception of the good economic activity is embedded within. This is a point that Polanyi makes on the very first page of his book.[53] The distinction is between a certain sort of conventional wisdom as embodied in theory and a pragmatic wisdom better reflected in actual practice. This theory–practice divide is echoed in Bruno Latour's claim that "we have never been modern," an argument that the theoretical concept of modernity is premised upon an impossible disembedding of society from a nature conceived as "other."[54] The reason this theory cannot be fully realized, Polanyi argues, is that it would require both humans and nature to be transformed in their entirety into something that they can never truly be: *commodities*.[55]

To fully commodify labor and land would be to reduce humans and nature to the quantities of labor activity or property that are exchanged in the market. Yet this reduction can never be anything more than a bizarre fiction.[56] Why? Commodities, Polanyi notes, are "objects produced for sale on the market," that is, human creations.[57] What particularly interests him, however, are what he terms "fictitious commodities"—labor, land, and money—which are obviously *not* produced for sale, therefore cannot be wholly subsumed to a market, and hence are not properly understood as commodities. Market liberal theory is thus dependent upon the impossible: labor can be bought but cannot be detached from the multifaceted lives of human subjects; land can be sold but cannot be detached from the complex of ecological relationships and assemblages within which it is situated. It may be illuminating to compare Polanyi's analysis of land with a more familiar environmentalist, Aldo Leopold. First, Polanyi: "What we call land is an element of nature inextricably interwoven with man's institutions. To isolate it and form a market out of it was perhaps the weirdest of all undertakings of our ancestors."[58] The "weirdness" of this commodified conception of land is echoed poetically by Leopold:

One hundred and twenty acres, according to the County Clerk, is the extent of my worldly domain. But the County Clerk is a sleepy fellow, who never looks at his record books before nine o'clock. What they would show at daybreak is the question here at issue . . . at daybreak . . . it is not only boundaries that disappear, but also the thought of being bounded. Expanses unknown to deed or map are known to every dawn, and solitude, supposed no longer to exist in my county, extends on every hand as far as the dew can reach.[59]

Material *effects* are central to this critique of the fictitious commodification of nature, but it is also clear here that this materiality cannot be disentangled from human ideas and *affects*. The critique highlights the fact that although land or nature cannot be produced for the market the former is a necessary precondition for the latter. Yet it is not simply material conditions that are neglected by fictitious commodification; vital aspects of human experience and *meaning* are also squeezed out. Fictitious commodities are hybrids: neither pure human construction nor pristine nature. This is an observation reiterated by a number of contemporary scholars.[60] Polanyi's particular insight here is that because nature is impossible to commodify fully for both these reasons efforts to do so inevitably create a basis for resistance—what he has termed the counter-movement—to liberalization and (false) commodification.

For Polanyi, the quixotic effort to fully disembed a private free-market system spontaneously generated a countermovement of resistance from among those whose interests, values, or livelihoods were threatened by the process of fictitious commodification. As he put it memorably, "laissez-faire was planned; planning was not."[61] These interests, values, and livelihoods are far more diverse than, say, any Marxist conception of class interest and reflect our myriad roles and experiences as "neighbors, professional persons, consumers, pedestrians, commuters, sportsmen, hikers, gardeners, patients, mothers, or lovers."[62] In this sense, Polanyi conceptualized resistance broadly as a "defense of man, nature, and productive organization."[63]

To conceptualize a countermovement in this manner is to present it as rooted in *democratic* impulses. That is, to argue that resistance has and can come from groups throughout society and that it is a reaction to the pervasive process of commodification is to contend that popular resistance is just that—popular. It is, Polanyi concludes, "actuated by a purely pragmatic spirit."[64] Again, Polanyi's countermovement is a manifestation of a Deweyan public, activated by the indirect consequences of actions otherwise construed as private. Moreover, each theorist offers a corrective to the other. Although Dewey himself has been criticized for the amorphousness of his notion of a public, as I've noted previously, Polanyi's account does much to flesh it out without misleading us into thinking it has a fixed institutional form. Conversely, although Polanyi imagines diverse sources of resistance he appears to downplay the plurality of ways in which we might interpret our interests, values, and livelihoods. This is most evident in the sense of inevitability or determinism with which he characterizes the countermovement. Dewey's emphasis upon the emergence of a public from these private roles highlights its contested quality and the lack of any fixed or assured outcome of such publicness.

The lack of an assured outcome of a countermovement against the (now neoliberal) quest for disembeddedness is also reflected in a key ambiguity in Polanyi's own work. What might count as an alternative to the utopian project of market liberalism? Polanyi's position is not clear. Mitchell Bernard describes the interpretive dilemma: "The counter-movement that Polanyi described was thus incapable of re-embedding economic life as Polanyi originally understood embedding, but it is not entirely clear from Polanyi's writing if he understood this as a contradiction-laden compromise or some sort of actual re-embedding."[65] To work through this ambiguity, we might return to the role that Polanyi prescribes for a conception of the good in structuring and guiding economic activity and the relationship of this conception of the good to environmental arguments. We know that this role offers a contrast to classical liberal *theory*, which conceives of itself as neutral with regard to the good. Thus, the state's role can be conceived in a minimalist manner, because it is said to keep a hands-off approach here. The boundaries of the political sphere are narrowly circumscribed. It is this proclaimed neutrality, we have seen, that is the basis for what Polanyi characterizes as the liberal project's utopianism. Yet it is the inability to realize this utopian ideal that ironically provides a basis for much of what Fred Block describes as its "intellectual resilience." In practice, societies will necessarily pull back from the full implementation of the market ideal. As a result, market advocates can always contend that their failure is a result of a "lack of political will" rather than a failure of their ideas.[66] Examples of this contention can be seen in the aftermath of the post-Soviet attempts at economic "shock therapy," the California energy deregulation debacle of the early 2000s, and the Republican response to the global financial crisis and recession of 2008.

Political practice simply cannot be devoid of any idea of the good. This is an ontological point from which we can draw important normative conclusions. First, if the size of the state or its planning functions is not, in practice, primarily a function of whether or not it is guided by a liberal philosophy, then advocating planning, regulation, or more far-reaching efforts to define the character of economic activity per se cannot be said to be identified with limitations on freedom. Second, if the implementation of market liberalism reflects the particular political agenda of activist states, then the claim to inevitability on behalf of this agenda is undermined and alternatives become more readily imaginable.[67]

The theory of market liberalism argues for very narrow boundaries for the political sphere, yet in practice these boundaries are necessarily quite broad. In the previous paragraph, I suggested that a recognition of this fact makes the sort of changes proposed by many environmentalists seem more viable, because such recognition doesn't require a complete transformation of either existing public philosophy or of the extant role of the state. Yet the task for philosophers, political theorists, environmentalists, and all concerned citizens appears more difficult from this perspective. We cannot rely, as at times Polanyi appears to, upon a simple argument that the economy must be "reembedded" in socioecological relations. It is not embeddedness, morality, or the good per se that must be defended here. Instead, the task of the critic must be to develop a particular case for a conception of the (environmental) good here and the institutions that can carry it forward. It's not *the* good, but *this* good; not abstract "morality," but *these* moral principles; not an "embedded" economy, but one embedded in a *particular* sort of socioecological structure, which must be advanced.

To Polanyi, the divide among critics of the unrealizable market-liberal utopia is a disagreement over whether "the idea of freedom can be upheld or not."[68] This idea of freedom is dependent upon an acceptance of the reality of society and hence the inescapable need for planning, regulation, and public control. Thus the problem is truly one of protecting what Polanyi terms "freedom in a complex society."[69] Whereas others characterize freedom as the abstraction of negative liberty, here freedom becomes intimately tied to the identification and pursuit of an environmentally relevant conception of the good. Here, as elsewhere, governance plays a necessary role. In structuring a process that allows for the public identification of the good, it need not abandon a commitment to freedom in private life and indeed can further its pursuit. Drawing insight from Polanyi, we can believe that such freedom is possible precisely because he offers hope that a democratic public will find it compatible with the pursuit of environmental sustainability.

II

Contesting the Material Practices of Everyday Life

5

Land and the Concept of Private Property

For all its technical complexity and inflexibility, for all its faults, it is still true that the law of property is one of the most important conversations that we Americans have with the natural world.

—Theodore Steinberg[1]

In 2004, 61 percent of voters in the state of Oregon passed Measure 37, a ballot initiative that supporters claimed would protect landowners from the government "taking" their rights through regulation on the development and use of their land. Where such regulation exists, this initiative required that owners receive either "just compensation" or an exemption from it.[2] Measure 37 proponents, like advocates of similar efforts across the US, were well aware that to provide this compensation would quickly bankrupt state and local governments; the measure was designed as a roadblock—and roll-back—to Oregon's widely noted land-use and environmental planning policies.

After passage, early reports concluded that "Measure 37 is starting to unravel smart-growth laws that have defined living patterns, set land prices and protected open space in this state for more than three decades."[3] Moreover, analysts at the time observed that as a result of Oregon's initiative "anti-sprawl legislation has lost political momentum across the country."[4] Yet several similar ballot initiatives subsequently lost in other western states.[5] Three years later, in November 2007, Oregonians passed Measure 49 by a similar margin, which reversed commercial and large residential development rights claimed under the earlier initiative. In the wake of these seemingly conflicted outcomes, then, Theodore Steinberg's observation about property law seems apt.

The notion that regulation constitutes a "taking" of property has been central to the so-called "property rights movement" across the United States and has emerged as a major challenge to land-use and other environmental legislation.[6] Although the movement is heavily funded by corporations with interests in resource extraction and property development, its messages have proven particularly salient among citizens for whom real property (most typically a single family home and the land it is built upon) is the dominant proportion of their wealth—largely middle- and lower-middle-income residents.[7] The movement taps into a familiar argument that owning property means that we have something of an absolute natural right to it—a right protected by legitimate government but one that should not be limited or modified by government. The property rights movement attributes a constitutional locus to this concept of property via an expansive reading of the so-called takings clause (also known as the eminent domain clause) at the end of the Fifth Amendment to the US Constitution: "nor shall private property be taken for public use without just compensation."

Although the 2004 Oregon vote is one instance that suggests that this concept of property can be quite salient even in a so-called "blue" state, the roll-back that resulted from the subsequent 2007 vote indicates that this is not the whole story. Even before the 2007 vote, observers had noted that there seems to be "a nationwide paradox in public opinion"; this concept of property in land generates popular support, but citizens also routinely endorse policies to protect themselves from environmental harms, to reduce the length of their daily commute, and to preserve open space and aesthetic vistas.[8] One policy observer summarizes the paradox by noting that "there is something puzzling about Measure 37's success. According to statewide surveys, Oregonians either support or strongly support land use regulations in almost the same proportions by which Measure 37 passed."[9]

All this must give pause to environmentalists. Even among some sympathetic constituencies, property can be viewed in ways antithetical to environmental policies. One way out of this paradox might be to blame it entirely on corporate special interests breeding false consciousness: they dupe people into supporting a measure that is against their real interests. Alternately, we might blame it on "the people": they want to have their cake and eat it too; they are confused and support contradictory goods. Such "outsider" criticisms are not wholly without merit. In this chapter, however, I argue that we can make better sense of these contradictory pressures by distinguishing private property in land from an absolutist concept of property ownership. This can allow us to recognize and accept the undeniable attraction of private property for many while also tracing the equally undeniable and to-be-anticipated sources of popular resistance to absolute ownership. To do so, we must unpack some of the core assumptions upon which the absolutist concept is erected, highlighting the ways in which they shape and constrain our understanding of land and other actual property. This will require distinguishing between theory and practice in contemporary societies such as the United States, and because of the central role of the law in both the theory and practice of property this chapter will focus upon US jurisprudential arguments far more than others in the book. I argue that the absolutist concept of property ownership is the manifestation of a liberal theory that can never be fully realized.[10] As such, it reflects a willful utopianism that is necessarily challenged by everyday experience and practice.[11]

One surprising reason that many environmental activists and theorists have trouble effectively responding to "takings" proponents is because they unintentionally reinforce the absolutist concept of what it means to own private property. This is *not* to say that they endorse this view of property ownership. Yet the absolutist concept often plays an unacknowledged yet decisive role in shaping their diagnosis of the environmental ills of contemporary society. By exploring their limitations, I set the stage for an alternative concept of property as inherently embedded in both social and ecological relations. Delineating this concept as it emerges from practice, I argue that the familiarity of these practices makes the concept more likely to resonate with substantial portions of the public. In this way, it offers greater democratic possibilities than other familiar environmental views discussed here. In this way, I seek to identify space *within* western practices of property for a critique of the move to commodify land to coalesce—space that might bolster efforts to reclaim popular support from the property rights movement advocates.

The chapter, then, proceeds as follows: In the first section, I delineate the absolutist concept of property. Although it should surprise no one that libertarian theorists, including Robert Nozick and Richard Epstein, are clear exponents of this concept, I draw upon a more diverse array of prominent property theorists—including A. M. Honoré and Jeremy Waldron—to establish its ubiquity. Only by recognizing this ubiquity can we comprehend the extent to which it closes off alternative ways of thinking about the scope and character of property ownership. In the second section, I illustrate the influence of this concept within several significant strands of contemporary environmental argument.

Some theorists have recognized and called attention to the inadequacy of the absolutist concept, describing it as a consequence of our changing practices of ownership. In the third section, I consider the analysis of two such exponents, legal theorist Thomas Grey and environmental philosopher Gary Varner. Grey and Varner are exemplary in their attention—respectively—to the social and ecological context of our property practices. Yet I challenge their temporal accounts of change; both mistakenly imagine a bygone era in which our practices actually did correspond with the absolutist concept.

What might it mean to place Grey and Varner's account of contemporary practices at the core of a reconceptualization of property itself? I explore this question in the fourth section by outlining a pragmatic concept of property as necessarily—not contingently—embedded in social and ecological relations. Such a concept may lack the parsimony of the absolutist concept, because it emerges out of the messy world of material practice itself, but it makes up for this in allowing us to make sense of a world in which private property is salient, yet society and ecology cannot—and should not—be ignored. In the final section, I argue that environmental criticism rooted in this pragmatic and embedded concept of property offers promise for a broader and more pluralistic environmental politics; I conclude by suggesting that some glimmers of this politics might be found in the success of Measure 49 in Oregon.

One final note: although the absolutist view of property ownership may be particularly salient in the United States, it is central to the globalist neoliberal economic agenda.[12] Thus, although the concept was once characterized as a manifestation of American exceptionalism, the globalist effort to flatten land tenure arrangements and other property relations makes this concept look more commonplace and less exceptional all the time. Still, the US case may be an especially difficult context within which to develop my argument persuasively. By doing so in this context, the confusion of theory and practice that I identify might help to address similar dilemmas elsewhere.

The Absolutist Concept of Property Ownership

The right of property [is] that sole and despotic dominion which one man claims and exercises over the external things of the world, in total exclusion of the right of any other individual in the universe.

—Sir William Blackstone, *Commentaries on the Laws of England, Book the Second, Chapter the First* (1765–1769)

In the United States, Blackstone's image of property as absolute dominion sounds familiar even to those who have never heard of its author or his *Commentaries*. As noted in the last chapter, "a man's home is his castle" is an aphorism taken seriously by many. It is also reflected in realtors' and lawyers' talk of "fee simple" or "freehold" ownership and in the notion of unified ownership captured in the legal maxim *cujus est solum ejus est usque ad coelum* (whoever owns the soil owns all the way to heaven).[13] This view of private property resonates widely. Yet the normative defense of this view is most explicitly advanced by libertarian activists and classical liberal theorists, so I will turn to their arguments first.

The absolutist concept regards private property as unitary, precluding legitimate restrictions on possession, use, or transmission of property.[14] It conceives of property as isolated from its context within a web of social and ecological relations that generate diverse claims of public interests and that might be in tension with the will of the property owner.[15]

Taken to its logical conclusion, this notion of property ownership would simultaneously prefigure and negate the concept of political sovereignty.[16] After all, if property is truly held as a unitary and exclusive right, then the right-holder is sovereign.[17] True believers in the absolutist concept thus have the benefit of consistency if they reject any legitimate role for the state in this regard. As libertarian Robert Nozick makes clear, it is the violation of this absolutist unity of property that agitates individualist anarchists. Although sympathetic, Nozick ultimately rejects their position in favor of a "minimal state" that protects property from incursion by other members of civil society.[18]

Rather than modify the concept of property ownership itself, however, the defense of a minimal state emerges only as a practical concession for libertarians such as Nozick. As a result, political sovereignty is conceived as a necessary evil that intervenes in and violates the preexisting private sphere of ownership.[19] Such intervention ("police power") is only tolerable to libertarians if it both protects preexisting property and ensures a proportional allocation of all gains that result from the security that this protection provides.

Richard Epstein seeks to anchor Nozick's libertarianism in constitutional law, tying it to the Fifth Amendment's takings clause. Earlier in US history, a "taking" under this provision was understood to result only from the physical seizure of land or other property by the government.[20] As a result, both the scope of the clause and the complexity of determining appropriate compensation were limited. In Epstein's expansive reading, however, virtually any regulation of, or so-called intervention into, private property right must result in compensation or it is illegitimate. In this manner, the clause becomes the twelve-word manifestation of an absolutist concept of property and, by extension, an underlying libertarian political philosophy. The power to act on behalf of a public good is recognized here, but (literally) only at a price: the "just compensation" of all private property holders. Legitimate government action requires the purchase of private property rights by the government for public use or benefit. These, in turn, become conceptualized as *public* property rights—the direct, unmediated exercise of sovereignty.[21]

Though tempting, I do not intend to try readers' patience with yet another criticism of these libertarian claims. Instead, I call attention to the ways in which the absolutist concept of property is reinforced by many who are not libertarians and who do *not* normatively embrace it. Joseph Singer has concluded that most philosophers and political theorists who discuss the issue—despite their awareness that "many conceptions of property are possible, and that limits on full ownership rights are pervasive" and despite deep disagreement "about how much regulation is appropriate"—treat the absolutist concept as the ideal-type.[22] This ideal-type helps explain why it is often so challenging to reply with conviction to takings proponents.

For example, legal theorists often cite A. M. Honoré's essay "Ownership" as one of the most widely respected philosophical discussions of property ownership. Honoré—no libertarian—appears to break down the unity of the absolutist concept by analyzing eleven distinct "incidents of ownership," including familiar rights to possess, use, and transmit property.[23] Modern legal systems frequently separate such rights from one another—allowing owners to sell or rent particular incidents and governments to restrict their exercise in the name of the public interest. At first glance, then, Honoré might appear to be a critic of the absolutist concept. Yet his analytical precision aims to further our understanding of the scope of that concept itself—which he terms "the 'liberal' concept of 'full' individual ownership."[24] Teasing apart the incidents of property ownership, for Honoré, allows us to identify deviations from the liberal or absolutist concept. Although Honoré is not a libertarian, then, he nonetheless accepts the absolutist concept as a regulative ideal, never using his analysis to rethink that concept itself.

Another example can be found in Jeremy Waldron's exhaustive study of the right to private property. Waldron, also no libertarian, reinforces Honoré's analysis by developing a distinction between the "concept" and a "conception" of private property. He describes the concept as the abstract "idea of ownership" itself, whereas the conception operates at the level of "the detailed rules of particular systems of private property." He confines the normative contest about property to the lower level of competing *conceptions* of it while—again—describing that the overarching *concept* of private property ownership in absolutist terms.[25]

Although they do not aim to advance the absolutist concept as a normative guide to property practice, scholars such as Honoré and Waldron nonetheless ease the way for those who do have this goal. From this vantage point, what Waldron terms *conceptions* of property to some degree must grapple with the existential web of social and ecological relations within which property exists. Yet these are viewed as deviations from the supposed ideal-type of the ownership *concept* itself. Conversely, normative arguments that seek to strengthen respect for property's embeddedness within this web are more difficult to legitimate than they might be otherwise. Here, government action on behalf of public interests is necessarily viewed as a restriction upon or intervention into property properly conceived.[26] Thus the burden of proof needed to justify such action is great, and the absence of such violations will be viewed as the default position whenever the burden of proof is not met.[27] It is striking, then, that even when some of the most methodical and insightful theorists recognize an incongruity between the absolutist concept and the actual practice of property they nonetheless reify and legitimize the idea of absolute ownership as central to their idea of private property.

There is, however, a further challenge for this position. We have seen that Honoré identified a number of distinct incidents of property ownership and that these also highlight possibilities for different individuals to hold different rights or "incidents" over the same piece of land or other property object.[28] Lawyers and law professors today take this sort of divided ownership for granted, arguing that "instead of defining the relationship between a person and 'his' things, property law discusses the relationships that arise *between people* with respect to things." In these relationships, different individuals (or corporate entities) are understood as capable of owning discrete "bundles of rights" with respect to some thing.[29] Here too a challenge to the coherence of the absolutist concept would appear to emerge. After all, if the concept relies upon a claim of unified ownership of a thing, then the lawyers seem to have debunked this. But the concept proves more resilient than this might suggest.

Consider air. If the absolutist regards ownership of a thing as unified, we should expect that the owner of a piece of land—real estate—would also be regarded as owning *usque ad coelum* (up to heaven). Such unity can be understood as physically violated by aircraft and regulated by a zoning law that restricts building height. These clearly modify the property right characterized in such a manner, though arguably not by all that much.[30] But what if we first disaggregate the property owner's "bundle of rights," recognizing that one property right in this bundle is to the air above his or her piece of land? The economic value of such "air rights"—especially in densely populated places like Manhattan—could then become enormous. Moreover, these air rights could be sold to someone other than the owner of the land. In this context, a regulation restricting building height can be viewed as an egregious violation—a "taking"—of an owner's property in the air.[31] More generally, although a government action may appear as a modest or even trivial intervention into an owner's property right understood as a singularity it can almost always appear to be a far more comprehensive—even absolute—violation of some particular incident, or right, within the owner's bundle.[32] We can return to Epstein to summarize the claim: "No matter how the basic entitlements contained within the bundle of ownership rights are divided and no matter how many times the division takes place, all the pieces together, and each of them individually, fall within the scope of the eminent domain [i.e., "takings"] clause."[33] In this way, the lawyer's "bundle of rights" view—which at first might seem to temper it—actually sets the stage for a far more expansive application of the absolutist concept of property.

When political theorists consider private property, often the first question addressed is the justification for ownership claims. Familiar justifications—rooted in first possession, labor, utility, convention, and personal development—have been traced to Locke, Bentham, Hume, Mill, and Hegel, among others. Yet justificatory arguments are intimately tied to questions about the legitimate *scope* of private property arrangements. The central question here is what "things are capable of being owned" and what ought to remain "outside commerce."[34] For example, debate about the commodification that results from patent rights to genes, the buying and selling of human organs, and intellectual property rights all reflect the contested scope of private property. Things outside the bounds of private property need not be viewed as alike. They might be viewed as state property or as part of "the commons."[35] They might be viewed not as objects at all but as subjects who are intrinsically valuable. Opposition to human slavery is a paradigmatic example of restraining the scope of private property in this way. Finally, to know that something is regarded as private property does not exhaust the question of what it *means* to own it—that is, the character of ownership. The absolutist concept offers one answer to this question, but not the only one.

Yet the absolutist concept can suppress other characterizations of ownership by presupposing its own answer to be uncontestable. A "commodity" has been defined as a thing that is produced for sale or exchange and so perfectly alienable—consistent with the absolutist concept.[36] In this way, the distinction between the scope and the character of private property is elided; the observation that an object is private property is regarded as synonymous with a claim that it has become property in the absolutist sense. By contrast, as Lewis Hyde has shown, property sometimes participates in a nonmarket system of gift exchange in which objects may be held privately but under a radically different set of conditions.[37] More generally, I will argue that the formulation of an alternative to the absolutist concept can begin from the recognition that property is rarely—and land is never—fully commodified.

Three Tendencies in Environmental Argument

The absolutist concept of property shapes the way environmentalists have conceptualized the role of ownership—especially of land—in society. As a result, it tends to constrain the character of the alternatives that they construct and point them in undemocratic directions. To be sure, there is a great diversity of environmental movements and ideas. Making no claim to be comprehensive, here I offer three ideal-typical environmental positions regarding property ownership—the first two drawn from the contrasting models of "inside players" and "outside critics" developed earlier in the book. The third is the position of so-called free-market environmentalists and libertarians, who diverge from the views of the others by enthusiastically promoting the absolutist concept of property. My consideration of each is structured by my argument that these otherwise diverse approaches are all shaped by the absolutist concept.

Uncontested Concept

What I have termed the environmentalist approach of "inside players" John Dryzek has termed "administrative rationalism" for its emphasis upon the role of expertise and the tools of the administrative and regulatory state to advance environmental protection.[38] In pursuing this path, the state "intervenes" in the economy and so the absolutist concept of private property and land ownership is seemingly "violated." Despite this active role for regulation and other government action in relation to property, however, the absolutist concept itself is left uncontested.

Because it is uncontested, these "insider" environmentalists typically do not have much to say about property per se. The goal seems to prevent violations of absolute ownership from becoming too provocative. In this way, regulations can be cast as sensible and modest efforts to manage problems, enabling environmentalists to dismiss critics as extremists due to their unwillingness to brook touching property rights at all. This approach, however, challenges neither the language of "intervention" or "violation" of private property nor the characterization of the regulatory state as an external imposition into an autonomous private sphere. As a result, regulations are regarded as a necessary evil: necessary because of the environmental good to be protected, yet evil because of their inherent violation of property conceptualized in an absolutist manner. This approach echoes Waldron's approach: advancing a distinct *conception* of property rights while leaving the overarching absolutist concept uncontested.

Although this approach seeks to *avoid* or minimize conflict with the absolutist concept, a variation seeks to *evade* the absolutist proscription on public action. The intent is to dissociate environmental regulation from other property-related issues by rooting the former in the objective analysis of ecological science. Here, land is highlighted as distinct from other types of property. Proponents can then appeal to the courts to implement this science objectively. Environmental lawyer David Hunter captures the essence of this approach: "Admittedly, protecting individual property holders from arbitrary majoritarian value choices is the fifth amendment's primary purpose. However, the need to preserve the ecological functions of certain types of land is not a majoritarian value choice but an ecological imperative, thus differentiating land-use restrictions on ecologically sensitive lands from other property restrictions."[39] For nonenvironmental policies, Hunter suggests that the appropriate question, understood in accordance with the absolutist concept, is whether a regulation has taken private property. He presumes that governments in the United States make regulatory decisions based upon "majoritarian value choices" and appears comfortable with a concept of property that blocks such subjective—"arbitrary"—choices. Only where ecological concerns regarding land predominate does Hunter argue differently. Here he claims that it is not truly the government that regulates private property; it is the land itself that deems the restrictions necessary. As a result, it is inappropriate to hold the government responsible for providing compensation.

Hunter does not challenge the absolutist concept but argues for a repositioning of ecological science *within* it. To make his case, he must claim that environmental protection decisions can be made on the basis of objective scientific imperatives and can thus be clearly differentiated from subjective political decisions ("majoritarian value choices"). He argues that one "reason justifying a move towards an environmentally sound property law is that we cannot afford to wait for social value changes to be reflected more boldly in the political process. Ecologists see the scientific imperative of their view as legitimating the imposition of new laws on a society that perhaps has not yet been steeped in the environmental sciences."[40] Hunter presumes that a defining feature of what I am terming the absolutist concept of property is a distinction between subjective, nonmaterial criteria for law and objective, material ones. Although the former violate property rights, the latter do not. His aim is not to move us from one side to the other or to challenge the distinction itself. It is, instead, to argue that scientific, ecological issues have been mislocated in the realm of subjectivism and ought to be repositioned on the other side—as a kind of objective, material harm. He further presumes that the judiciary is willing and able to implement directives drawn from ecological science objectively, whereas more democratic bodies fail to do so.

Transformative Change

Whereas the previous environmentalist approach seeks to avoid conflict with the absolutist concept of property, some radical environmentalists act as outside critics and therefore reject private property (while still accepting the absolutist concept as an appropriate description of property's meaning). One version of this approach is ecosocialism. Here, as among socialists more generally, the justification for private property ownership is rejected. The grounds for this rejection are, at least in part, said to be the absolutist character of this ownership—a character that excludes use-value in favor of exchange-value. Although some argue for replacing private ownership with a public, collective ownership that appears as its mirror image, others assert that "it is plain *hubris* to think that the earth, or nature, can be owned."[41] In either case, the rejection of private property is premised upon the conviction that the absolutist concept is an accurate description of contemporary land ownership arrangements.

A second version of the outsider environmental approach proposes ethical transformation as the basis for the redefinition of property. Aldo Leopold's well-known analysis of "The Land Ethic" in *A Sand County*

Almanac offers an apt model of this. Lamenting that "there is as yet no ethic dealing with man's relation to land," he identifies the absolutist concept of property as the core problem to be overcome by the introduction of such an ethic. In his words, "land . . . is still property." As property, "the land-relation is still strictly economic, entailing privileges but not obligations."[42]

In his quest to overcome this property concept, Leopold describes a paradox. Concurrent with the "growth in knowledge of land" and "good intentions toward land," he argues, has been the expanded "abuse of land."[43] The attempt to understand the causes of this paradox led him to describe two dichotomous world-views. He labeled this dichotomy, in an uncommonly prosaic moment, the "A–B Cleavage." Whereas "As" regard land merely as functional for "commodity-production," "Bs" regard its "function as something broader" and exhibit "the stirrings of an ecological conscience."[44] The first view holds that land is valuable only in a narrowly instrumental manner—as property subject to absolute ownership. Indeed, he asserts—and more recent writers continue to affirm—that most Americans adhere to this view of property.[45] The second view can see far greater and broader value in the nonhuman world as well. One may arrive at the latter view, Leopold argues, only by first expanding the horizons of one's "moral community" to include the land itself.[46] With the expansion of the moral community to encompass the land, Leopold argues that we will "change . . . the role of Homo sapiens from conqueror of the land-community to plain member and citizen of it. It implies respect for his fellow-members, and also respect for the community as such."[47]

By accepting this as true, Leopold and others conclude that the only avenue for an effective response to policies that emanate from this dominant view lies in the importation from the outside of a new and radically distinct ("B") view—the "land ethic"—to transform contemporary environmental attitudes and hence property practices.

Advocating Absolute Ownership

Although neither previous approach normatively promotes the absolutist concept, both take it to be an accurate description of property and seek to address the constraints it imposes upon contemporary discourse. One environmental approach, however, does advocate property absolutism.

Garrett Hardin's singularly influential 1968 essay "The Tragedy of the Commons" is often remembered for his countenance of coercion, yet he also described "selling off [the commons] as private property" as a means of avoiding "tragedy."[48] This approach has been advanced by so-called free market and libertarian environmentalists, as well as a growing number of legal scholars working on environmental topics.[49]

Unlike either of the previous approaches, this one is distinguished by the recognition (accurate, I argue) that an absolutist concept is actually at odds with present land ownership and other environmental practices. As Terry Anderson and Donald Leal put it, "the key to . . . free market environmentalism is to *establish* property rights that are well defined, enforced, and transferable."[50] Similarly, libertarian Tom Bethell argues that "the underlying cause" of such environmental crises as rainforest destruction "has been precisely the difficulty of obtaining clear title."[51] These authors argue that specifying an absolute private property right ensures that the owner has a self-interest in protecting their possession. This can be either because it protects long-term economic value or because harm to neighboring property becomes a trespass or tort, subject to legal injunction or compensation.[52]

Clearly, this approach seeks to expand the scope of private property into presently noncommodified realms. Anderson and Leal promote, for example, "property rights in free roaming wildlife," "the establishment of full property rights" to fishing sites, and "property rights to both waste and the disposal medium [e.g., air and water]."[53] Less obvious but equally important is that they seek to move what it means to own property in a more absolutist direction, rejecting the legitimacy of government regulation in favor of the common law tort system as a means of preventing destructive practices. Here, for example, libertarian Tibor Machan argues that "capitalism requires that pollution be punished as a legal offense that violates individual rights." This commits him "to resisting pollution regardless of the expense and inconvenience involved to those who pollute."[54]

The embrace of private property rights outlined here exasperates and disorients many other environmentalists. The exasperation often reflects skepticism about the sincerity of these proponents' environmental commitments. Mark Sagoff seeks to test this, pointing out that taking the inviolability of exclusive private property rights seriously would allow individuals to seek injunctive relief against polluters, thus enabling "individuals who refuse to be bought off" to "close the economy down."[55] Such an extreme conclusion, Sagoff argues, is self-defeating. Some sort of stopping point is needed, one that cannot be derived from the private property rights themselves. "Otherwise," Sagoff concludes, "the best will become an implacable enemy of the good and incremental progress toward pollution control will be rejected in view of an ideal of total purity that can never be achieved."[56] The disorientation is a reflection of the fact that this position, as its advocates themselves acknowledge, inverts the more familiar environmentalist views canvassed in my previous two subsections.[57] Where those reject or evade an absolutist concept of ownership that they take to be well established in practice, these libertarians and free marketeers embrace the same view, which they find to be *absent* from practice. Thus they wish to *impose* a dichotomy upon recalcitrant practices. After all, efforts to enshrine regulation as a "taking" of private property—like Oregon's Measure 37—are only consequential to the extent that land ownership is *not* now consistent with this concept of property.

In conclusion, the frequent elision of property ownership per se with the absolutist's concept of what this ownership entails is a mistake shared by otherwise radically diverse environmentalist positions. Where they identify inadequacies in the absolutist concept, this elision leads them to critique or reject private ownership per se. Ironically, it is only those who embrace the absolutist concept as a normative ideal who call our attention to its distance from actual practice.

The Failed Search for Absolute Property Ownership in Practice

Although the common idea of property may, by its nature, be the most absolute of rights, the institution of property must, by its nature, be the most compromised.
—Laura S. Underkuffler-Freund[58]

To conceptualize property in absolutist terms is both pervasive and wrong. So far my goal has been to establish its pervasiveness; here I develop the latter claim. A danger in developing this argument is that readers may suspect that I am setting up a straw man. Surely the absolutist concept is widely recognized as inconsistent with actual practice? If so, then it would not seem to take much to establish its conceptual inadequacies as well. Yet even critics whose subject is the inconsistency of absolutist property with practice appear to have difficulty conceptualizing property in another way.

In this section, I examine two notable and incisive essays on this subject, both with evocative titles. Thomas Grey, a legal theorist, entitled his "The Disintegration of Property."[59] Gary Varner, an environmental philosopher, wrote "Environmental Law and the Eclipse of Land as Private Property."[60] Both essays are valuable for the ways they highlight the disparity between property theory and contemporary practices. At the same time, the rhetoric of disintegration and eclipse manifest in the titles is driven by their shared inability to imagine an alternative to the absolutist concept of property. The reason this can be so difficult, I suggest, is that the alternative—a concept of property as necessarily socially and ecologically embedded—cannot be formulated a priori but only emerges out of the entanglements of material practices themselves.

Grey argues that the twentieth century saw the rise of the lawyer's bundle of rights view and its divergence from the "ordinary" concept of property as "*things* that are *owned* by *persons*." The consequence has been "to dissolve the notion of ownership and to eliminate any necessary connection between property rights and things."[61] In Grey's telling, this divergence is both conceptual and historical. In the eighteenth century, "at the high point of classical liberal thought," Grey maintains that the absolutist concept was "at the center of the conceptual scheme of lawyers and political theorists." Key to his argument is the claim of congruence between this theory and then-contemporary practice: "It is not difficult to see how the idea of simple ownership came to dominate classical liberal legal and political thought . . . this conception of property mirrored economic reality to a much greater extent than it did before or has since. Much of the wealth of the preindustrial capitalist economy consisted of the houses and lots of freeholders, the land of peasant proprietors or small farmers, and the shops and tools of artisans."[62] Thus, as industrialized societies developed increasingly complex and abstract forms of property relations in practice, Grey maintains, "property" went from being "a central idea mirroring a clearly understood institution" to something that is "no longer a coherent or crucial category in our conceptual scheme. The concept of property and the institution of property have disintegrated."[63]

Varner offers a similar historical and conceptual account, albeit from a different vantage point. Focusing upon the growth of environmental law and regulation, he argues that "the greater the number of sticks government removes from the bundle of rights landowners hold, the closer it comes to eclipsing land as private property."[64] To say this, of course, presumes that we first conceptualize private property in absolutist terms as holding all the sticks in the bundle. Varner makes this clear, describing property ownership as "being legally permitted to do with it as one will, independently of others' wishes."[65] Whereas Grey's account focuses our attention upon the move away from preindustrial forms of property, such as land, Varner describes a change taking place *within* landed property. The change, in his account, took place "in the final third of the twentieth century" in which "environmental laws and regulations proliferate[d]."[66] He emphasizes the ecological interconnections between and among land uses and argues that environmental regulation is properly understood as an exercise of the state's police power, intending to prevent harm to others, rather than an exercise of eminent domain power (which, in the United States, can only take property for a public use if compensation is provided):

My conclusion is that the eclipse of land as private property is near at hand . . . we have discovered that land uses depend so heavily on an ecological infrastructure—on processes that, if they are property at all, are inherently public property—that it hardly makes sense to conceive of land as private property. The proliferation of environmental regulation thus threatens to wrest unilateral control of land from the landowner to such an extent that the appellation 'private property' no longer naturally applies.[67]

Two distinct claims appear in both Grey and Varner's accounts. The first is a temporal claim about the demise of property: once upon a time, the (absolutist) concept of private property fit reasonably well with actual practice, but it no longer does. The second is a claim that the present ill fit results from the growing recognition of property's embeddedness. This highlights the gulf between the theory and practice of property. Grey focuses upon property's practical location within a complex web of social relations, whereas Varner focuses upon its ecological context. The absolutist concept of property necessarily excludes both these forms of embeddedness, which are appropriately viewed as mani-festations of material practices.

The temporal arc perceived by Grey and Varner explains the language of "disintegration" and "eclipse" in their titles. It is worth noting that neither of these theorists laments property's supposed demise; in fact both find good reasons for it. Nonetheless, I challenge this temporal account, arguing that in practice property ownership was never absolute, because even in principle it is never possible to disembed it from all social and ecological relations. The corollary is that the embedded or "mixed" property that Grey and Varner convincingly identify today was always already present.

Grey's and Varner's accounts both hearken back to a stylized image of an era in which absolutist property existed. When might this have been? Varner unconvincingly places it prior to the last third of the twen-tieth century, but this might only work if we keep the focus narrowly on the particular type of environmental regulation that emerged in the early 1970s. Property ownership was certainly encumbered before that. The existence of significant regulation of property is widely recognized throughout the twentieth century; the only serious question might be whether it played a significant role in earlier eras. Grey's answer, locat-ing practices of absolute property in the early modern era, may seem more plausible. Clearly, we cannot go back further into European history or we will return to the complex and constrained forms of land tenure that characterized feudalism.[68] Yet Colonial America is said to have escaped the European feudal legacy;[69] perhaps we could locate the abso-lutist concept of property there?

In William Cronon's insightful account of the ecological transforma-tion of New England as it passed from Indian to Colonial land tenure, he does conclude that "it was the treatment of land and property as com-modities traded at market that distinguished English concepts of owner-ship from Indian ones."[70] Yet although he rightly draws this contrast he is also right to note that commodification was far from complete. Indeed, although the colonists sought to create "permanent private rights" to property "these rights were never absolute, since both town and colony retained sovereignty and could impose a variety of restrictions on how land might be used. Burning might be prohibited on it during certain seasons . . . [property claims] might be contingent on the land being used for a specific purpose—such as the building of a mill—and there was initially a requirement in Massachusetts that all land be improved within three years or its owners would forfeit rights to it."[71] In sum, as severe as the contrast between Indian and European views of the land was, neither can be said to adhere to an absolutist concept of property in it. Thus the seventeenth century does not seem a good fit for Grey's thesis.

Perhaps the eighteenth century will do? After all, Morton Horowitz's influential account of American law—echoing Blackstone—states that "in the eighteenth century, the right to property had been the right to absolute dominion over land."[72] Yet this eighteenth-century right is not what it may first appear to be. Although it did provide a large amount of protection for property "to be enjoyed for its own sake," it also "conferred on an owner the power to prevent any use of his neighbor's land that conflicted with his own quiet enjoyment."[73] Two common law principles shaped the concept of property in this society: "the limiting do-no-harm rule of *sic utere tuo* and the more overarching moral principle of *salus populi suprema lex est*—the welfare of the people is the supreme law."[74] From these, concludes historian William Novak, "flowed a multitude of governmental restrictions on property, contract, morality, and a host of other aspects of social life."[75] Absolute dominion thus carried with it a severe antidevelopmental bias that continued into the mid-nineteenth century.[76] Indeed, because of recent legal and historical scholarship "we now know just how heavily regulated property was in pre-nineteenth century America."[77] This research supports the conclusion that "the use of land was being regulated—often very severely regulated—throughout English and early American history."[78] In sum, the eighteenth century's notion of "absolute dominion" is quite different from the absolutist concept. Significantly, although the Fifth Amendment language on taking private property was adopted during this era it was not until the twentieth century that some—including advocates of Oregon's Measure 37—came to argue that regulation and other government action could constitute a "taking" of property rights.

It is certainly true that later in the nineteenth century American law increasingly shifted to a prodevelopment stance, thus allowing much more aggressive and industrialized use of land.[79] Yet the history to this point makes clear the difficulty in ascribing this stance to an absolutist concept of property. Just as one owner's "quiet enjoyment" constrained the uses to which a neighboring owner might have put his property before this shift, so can the new stance constrain the first owner's enjoyment by enabling a neighbor to develop aggressively. Because property is embedded in social relations, decisions favoring one or another owner can be made, and decisions that balance rights among owners can be made, but none can favor the right of all to exercise supposedly absolute property rights.[80]

In sum, it is often recognized that the absolutist concept does not match contemporary property practices. But many—like Grey and Varner—presume that the concept did at least once correspond to practice. Having briefly canvassed the seventeenth through the twentieth centuries, we can conclude that the historical demise described by Grey and Varner is a fiction, because the absolutist concept that they posit as a historical artifact never existed. As such, the basis for this concept is called into question. Although committed libertarians may continue to hold it up as a normative ideal, the rest of us are better off reconceptualizing property as a necessarily hybrid concept that cannot escape its social and ecological context.[81] Grey and Varner recognize this hybridity in practice today, but their historical presuppositions prevented them from viewing it as central to property per se.

Embedded Concept (Not Just Conception) of Property

"Private property" cannot be defined or discovered a priori. This is the absolutist mistake. Property is created through social and material practices, and the conflicting claims that result from this impure starting point are typically resolved through governmental institutions, especially the courts. "Property rights," in Daniel Bromley's apt words, "are made, not found."[82] This pragmatic approach to property rights necessarily limits the precision with which we can delineate an alternative to the absolutist concept. Yet placing social and ecological embeddedness at the core of our understanding of property—rather than marginalizing them as deviations from or limits upon the idea of property—challenges two dichotomies central to the absolutist concept.

The first of these is between "private" and "public," the second between "artifice" and "nature." These categories—relevant though they are—can never be isolated from each other. Private property is intimately tied to public aims and combines the artifice of human labor and creativity with the broader material world. As my historical discussion makes clear, this reconceptualization of property is—first—a descriptive claim regarding what property always already is. Here I suggest the normative attractiveness of this concept as well.

Evidence of the public–private dialectic within property can be seen by returning to the prototypical legal understanding of property rights as "sticks in a bundle"—a disaggregable and changeable collection rather than a unitary and near-sovereign object. This view is meant to be purely empirical and, indeed, "scientific."[83] Yet as I described previously, Epstein and other takings proponents are able to incorporate this into the absolutist concept with surprising facility. Conversely, this understanding can prefigure a recognition of the inescapable role for public judgment in defining private property rights.

The public accommodation laws that emerged from the 1960s US civil rights movement offer a clear example of this. As Joseph Singer notes, "the idea that property should be open to the public without regard to race is wholly foreign to the understanding of the right of exclusion that prevailed . . . before the 1960s . . . The effect of this principle is to transfer a large stick in the bundle of property rights from the owner [of a business] to the public at large."[84] Singer makes the case for the revolutionary significance of this change, especially given the virulent opposition. Yet this change reflects a judgment about competing property claims rather than a diminution of an absolutist one. After all, as Singer goes on to argue, "the rules governing the property system not only must protect the interests of those who already own property, but also must establish legitimate conditions under which nonowners can acquire property rights." Here the customer's right to access and purchase goods or services clashes with the business owner's right to exclude. Thus, "the tension is *within* our idea of property itself," denying the absolutist presupposition that this is a case of public intervention into private property rights.[85] By contrast, Singer helps us to see that this is a case of conflicting rights in which either public action or inaction involves taking a stance in this clash. The tension cannot be escaped here, because it reflects property's inescapable embeddedness in social relations.

Although recognizing the publicness of private property is therefore vitally important, it does not directly advance ecological concerns. In fact, the first step in the lawyer's disaggregated view of private property is to emphasize the relationality of property between persons at the expense of the layman's view of property as "things," such as land.[86] In this way, it may seem to detach property from the biophysical world, suggesting a view of property as pure sociality. In order to avoid this conclusion, we must simultaneously address the relation between artifice ("pure sociality") and nature ("biophysical world").

Note that lawyers, planners, and other practitioners commonly deploy two related metaphors about property. The first of these is of property *rights* as "sticks in a bundle." The second describes property as a "bundle of rights." These phrases are often used interchangeably, so there is no particular reason to suspect that there is a tension between them. Yet I wish to draw upon these metaphors to develop a relevant distinction. The latter, but not the former, encourages us to see each particular right *as* property. This bundle of rights metaphor, as we have seen, reinforces Epstein's far reaching claim that each right in a property—individually and collectively—is subject to the constraints of the takings clause.[87] Thus a land-use law that prevents the development or use of land in a certain way is said to "take" that property known as a development right from its recognized owner—something that is legitimate only with "just compensation." If the government prevents me from developing a shopping mall on my farmland, for example, then they can be said to have taken my property, where this property is defined as the right to develop the land in this way. Here tangible or material property itself seems to vanish, to be replaced by a concept of the particular development or use right *as* property. By contrast, when describing property rights as "sticks in a bundle," each "stick" can be understood in reference to some identifiable thing, posing greater obstacles to this expansive interpretation. From this perspective, the property I own is the farmland itself—the bundle—and I would have to lose a great many individual sticks before I could meaningfully claim that the bundle had been taken from me.

Leif Wenar has argued persuasively that only by distinguishing property as an object, on the one hand, from rights to property, on the other, can we reconcile popular understandings of property with the insights of the legal community. By doing so, we sustain our recognition that the objects of property have an independent existence while we also recognize that the rights to this property are multiple and divisible. I argue that this is better captured by the sticks in a bundle metaphor, which highlights the multiplicity and divisibility of rights, and other normative obligations, *to* property. By contrast, the property as a bundle of rights metaphor seemingly prompts us also to view property as nothing more than the sum of the many rights claims themselves. It is more compelling to say, with Wenar, that "there are things that are property but that may not be *owned by* or do not *belong to* a single person."[88] In this way, we allow property to retain its lay connection to the material world while rejecting the naïveté of those who would assume that such connection requires an absolutist concept of property. Ecological embeddedness is integral to the very existence of what we call land; as such it must also become integral to our concept of property.

Embedded Concept of Property Prefigures a More Robust Environmentalism

What we call land is an element of nature inextricably interwoven with man's institutions. To isolate it and form a market out of it was perhaps the weirdest of all undertakings of our ancestors.
—Karl Polanyi, *The Great Transformation*[89]

Environmentalist argument can be bolstered by reconceptualizing private property as inherently—not just contingently—embedded within a web of social and ecological relationships. This has been the thrust of my argument. Here I wish to develop the claim further, noting some of the differences between such an environmentalist position and the three more familiar ones canvassed in the second section.

Unlike radical environmentalists who dismiss private property as necessarily absolutist and so must position themselves outside of a posited social consensus by rejecting it outright, we have seen that an environmentalist critic who embraces the embedded concept of property can appeal to a wealth of practices immanent to the society they wish to change. Unlike the mainstream environmentalist "player" who avoids or evades discussion of private property because they fear that an explicit challenge to the absolutist concept would weaken their appeal to a broad constituency, this environmentalist critic can remain outspoken and resonant in his or her rejection of a concept of property that can be convincingly shown to bear little resemblance to the actual property that these constituencies hold dear.

There is reason to suspect that the more recent victory of Measure 49 in Oregon indicates the promise of this sort of inside environmentalist criticism. In that campaign, proponents highlighted the ways in which the earlier initiative (Measure 37)—although touted as protecting the ability of private landowners to build a home on rural lands—was simultaneously used to justify thousands of claims for large-scale housing subdivisions and commercial development.[90] During its brief life, the earlier initiative generated over seven thousand property rights claims, demanding waivers for over $17 billion in government compensation.[91] Had the subdivisions and commercial development allowed under the measure proceeded, it would have inescapably entailed social impacts on other property owners and ecological impacts on sensitive lands and waterways. As one local activist in the scenic Hood River Valley put it, "It would have altered the valley forever."[92] Moreover, rather than create a level playing field, it introduced new inequities into property relations. For example, if a long-time land owner is granted development rights and builds a housing subdivision in what had been an exclusive farm-use zone, then neighbors lose, both in terms of the aesthetics and experience of farming and in terms of the need to negotiate frequent residential–agricultural conflicts over noise, smell, and other incompatibilities of use. Moreover, for those who purchased their land after the zoning had gone into effect in the 1970s based upon the expectation that the area would be exclusively agricultural, no legal claim under the measure could be made.[93] This difficult reality became increasingly apparent to residents as claims were publicized and adjudicated.[94] In sum, rather than clarifying and enforcing the boundaries and rights of absolute private property, the earlier initiative necessarily fostered a confusing array of complex legal, social, and ecological problems.[95] A countermovement of the affected, working against the assertion of property absolutism, is a thread that can be traced through Oregon's recent history.

A democratic movement built upon this sort of engaged, yet unapologetic, criticism is much needed among environmentalists today. An understanding of property as privately owned yet accountable to public environmental concerns and consequences is also much needed. The notion that the two can be mutually reinforcing has been the central burden of my argument in this chapter.

My goal has been to identify and explore democratic resources for a critique of the absolutist concept of private property ownership, which has been central to antienvironmental property initiatives such as Measure 37. Evidence of popular support for this initiative encouraged me to explore just how pervasive and problematic the absolutist concept is to our understandings of property. Although some may wish to blame this concept on public ignorance or wish to marginalize this concept as the purview of a narrow group of libertarian theorists, I have found that it shapes the understanding of property held by some of our most sophisticated thinkers on the subject. Although the conclusions I am able to draw are tentative, I hope to have shown that the democratic resources are not negligible.

The absolutist concept of property ownership will always be inconsistent with practice. Public delineation of appropriate uses for property is not new—although the absolutist concept fosters the illusion that it is—nor can it be eliminated, even in theory. Over time, the state has been the source of both efforts to impose (never fully successfully) the absolutist concept upon a recalcitrant practice and the institutional structure through which resistance movements have tempered its violence. As a result, efforts to impose "takings" restrictions will be far messier and more likely to encounter resistance than we would otherwise be led to expect.

Finally, I have argued that many environmentalist strategies and criticisms attempt to work from outside a posited social consensus. At the time, the electoral success of Measure 37 may seem to have confirmed their instincts in this regard, yet there are greater resources for social criticism than these strategies lead us to believe. Although such resources do not always cut so deeply, they can offer possibilities for expansive alliances among those with diverse identities and concerns—those threatened by efforts to impose an absolutist concept of property. In this way, they might provide a basis for a more expansive vision of active citizenship in relation to property concerns than the narrowly circumscribed role envisioned within the absolutist concept. Possibilities are not guarantees, of course, but in a period in which political hope is itself so elusive and the stakes are so high we would be foolish to ignore or dismiss them.

6

Automobility and Freedom

To speak, as people often do, of the "impact" of . . . the automobile upon society makes little more sense, by now, than to speak of the impact of the bone structure on the human body.
—Leo Marx[1]

Cars are central to modern life the world over.[2] When ownership and use of cars becomes widespread in a society—decades ago in most of the western world, more recently in China, India, and many other countries—we can identify substantial change: its centrality to economic activity and growth, transformation of land use and community development patterns, alterations to the rhythms of daily life, novel threats to individual safety, and of course its devastating effect on climate and environmental quality are only among the most evident.

The significance of cars in contemporary societies cannot be understood if our attention is focused narrowly upon the vehicles themselves. They must be recognized as a central embedded component in a material practice that many have come to call "automobility," an inclusive term that encompasses the roads and highways, parking structures, driveways and garages, traffic laws and enforcement, gas stations, refineries, dealerships and manufacturers, transformed urban, suburban, and rural forms and landscapes, and many other material components that are integral to driving an individual automobile. But automobility signifies more than this; also essential are imagery and attitudes toward driving and car culture, perceptions of space and speed and of the relation between technological innovation and cultural change, and the ways these intersect with social relations of gender, race, and class, as well as political discourse.[3] The material practice of automobility, in other words, is inextricably infused with cultural attitudes, associations, and perceptions. Although analytically separable, the physical and cultural aspects are entwined in automobility itself. It is because automobility is absolutely integral to modern societies that Leo Marx's epigraph is so apt. There is a growing literature on it by sociologists, historians, scholars of cultural studies, transportation planners, and a few political scientists. Yet automobility has rarely been the explicit subject of political theory.[4]

The entanglement of motorized vehicles and individual freedom can be found in the very etymology of "automobile"; the *Oxford English Dictionary* identifies the earliest definition as an adjective meaning "self-moving." It seems noteworthy that the "self," here, might be aptly understood as applied *either* to the vehicle or to the driver. My premise in this chapter is that the material practice of automobility is integral to our contemporary conceptions of individual freedom, and so reflection on automobility can and should inform critical discussions of this freedom. Familiar liberal and republican conceptions remain abstracted from practice in ways that make it difficult to garner insight into automobility. As a result, rather than beginning with such theories and then seeking to apply them to automobility, my aim is to examine automobility and freedom together. In what ways does the ubiquitous practice of automobility—one in which we participate whether we drive or not—shape understandings of individual freedom, and how does this understanding create both constraints and opportunities for critical evaluation of automobility itself? I leave it to others to contest and sort out whether and how the sort of discussion of automobility and freedom pursued here might reengage with established theoretical discourse on the latter.

On the one hand, automobility generates a tremendously flexible—and often appealing—source of mobility, privacy, and independence. On the other hand, it can be understood as a coercive practice that consumes massive amounts of space, requires lengthy commutes, increases dependence among youth, elderly, and others unable to drive or without access to a vehicle, relies upon extensive state surveillance, harms or eliminates environmental options for future generations, and structures patterns of living, working, and playing that often preclude many from more than nominal use of alternatives even where these are available. I seek to take seriously both these perspectives on freedom in turn here.

The View from "Autofreedom"

The cowboy spirit is about freedom, about going places and about answering to no one. The automobile not only embodies that spirit, it gives it life.
—Matt DeLorenzo, editor-in-chief, *Road and Track* magazine[5]

My aim in this section is to sketch, fairly, four senses in which automobility enables individual freedom. I do not present these arguments as inherently convincing or shared by all, but I also do not intend to construct straw man arguments. Although I raise crucial challenges for these manifestations of freedom in a later section, the challenges cannot be adequately appreciated unless we first take these manifestations seriously. That is, the challenge of developing resonant, engaged criticisms requires that we first truly recognize these influential, if often underarticulated, freedoms.

To take them seriously means that we cannot simply position ourselves as outsiders and dismiss them as "false consciousness."[6] Conversely, although I will have little to say about it here, nothing in this section is meant to naturalize this dominant view of freedom. That is, nothing here ought be viewed as inherently inconsistent with an empirical account of the rise of automobility's dominance that describes a heavy hand of the state and capitalist interests in structuring a particular model of economic development and of the built environment that reinforces this view of autofreedom, nor is it inconsistent to describe the multiplicity of ways in which popular culture reinforces this view.[7] I label four aspects of autofreedom that I consider here: identity, control, market preferences, and human flourishing.

Identity

Not too long ago, I whiled away a couple of hours waiting in a state government office, reading an especially critical and persuasive book on the politics of automobility. And yet, the experience couldn't have been more disorienting: the office was the Department of Motor Vehicles; I was there waiting for my sixteen-year-old son to take, and ultimately pass, his driver's license exam. Persuasive though it was, nothing in this book could counter Jake's enthusiasm: for him, as for many US teens in the past several generations, obtaining a driver's license is a vitally important rite of passage. Although as adults we might readily overlook or dismiss it, the exam itself requires a demonstration of knowledge and skill and—in a sense—an independent evaluation of maturity; passage reflects a mastery of complex and sometimes arcane state laws, as well as demonstration of driving skill witnessed and evaluated by agents of the state. The license reflects a key step toward full, adult involvement in society—and a substantial means of escape from parental surveillance. A license to drive wasn't the only thing that Jake wanted. To truly obtain the independence he had sought required, in his opinion, not merely a license but a vehicle—something that took much longer. If driving has been viewed as a key form of social participation, then the converse also seems true: Chella Rajan argues that "anyone incapable of owning and driving a car in present day North America has to be seen as lacking all the capacities and capabilities of citizenship."[8]

As a driver, in the words of Cotten Seiler, one has "opportunities for the spectacular expression of freedom and autonomy so affirming to the individualist,"[9] and that freedom is a central form of adult participation in many countries beyond North America. Even in relatively compact European countries with extensive public transit, the connection between driving and freedom is frequently made. Danish mobility scholar Malene Freudendal-Pedersen summarizes her empirical research: "When the interviewees were each asked why they have a car, the answers given had the same theme: 'I love the feeling of freedom' or 'it is simply the freedom and the time you save' or 'it gives me so much extra freedom to have the car' or 'when you're 18 you almost live in your car. We could do things—it was freedom.'"[10]

Yet although a license is a form of social involvement and driving may be seen as a capability of citizenship, participation in the system of auto-mobility is not manifest as a form of public engagement. Instead, as Seiler also notes, it is manifest as withdrawal from politics, "oriented toward 'a display of energy'—movement and consumption in lieu of democratic entanglement."[11]

Cars themselves have also been important forms of self-expression and expression of group identity. We've come a long way from Henry Ford's proclamation that "any customer can have a car painted any colour that he wants so long as it is black."[12] Manufacturers, of course, have become increasingly sophisticated in producing an array of colors, sizes, and styles that are advertised and marketed to distinct demographic identities and embraced by many as an expression of such identities. In addition, a history of car subcultures engaged in detailing and customizing vehicles has existed at least since the 1960s: "largely white and working-class 'hot rod' or 'stock car' enthusiasts and Chicano/Latino and African American 'lowriders,' as well as the more recent Asian-American 'import street racer' culture, illustrate the dominance of automobility even in the fashioning of distinctive 'ethnopolitical identit[ies].'"[13]

Control

Cars offer seemingly substantial flexibility, privacy, and control—all closely tied to notions of individual freedom. To drive is to not be bound to a train or bus schedule, or to the fixed routes that these vehicles travel on. Unlike walking or bicycling, which also allow for this flexibility, driving insulates one from the weather and often (but definitely not always) allows for movement across greater distances more quickly. Driving can allow me to go where I want, when I want to—day or night, summer or winter, rain or shine. Perhaps it is not surprising, then, that at least in the United States it is "the most automobile-like public transit mode, taxicabs, [that] already carries more passengers than all other kinds of public transit . . . put together."[14]

Cars allow us to control our interior environment—both to keep out the external elements and to adjust temperature and seating to our preferences. Driving also allows for substantial control over where we live, work, shop, and play. Again, one of Freudendal-Pedersen's Danish informants is valuable for identifying this as more than North American exceptionalism: "Cars take you exactly where you want to go. You can take a detour and then drive 3km down a road, and jump out on a deserted beach, where there is no one else. One feels a certain power when driving a car: this is why people love it—power and freedom."[15]

Loren Lomasky—in a philosophical defense of driving—summarizes this element of control hyperbolically yet with considerable insight:

In the latter part of the twentieth century, being a self-mover entails, to a significant extent, being a motorist. Because we have cars we can, more than any other people in history, choose where we will live and where we will work, and separate these two choices from each other. We can more easily avail ourselves of near and distant pleasures, at a schedule tailored to individual preference. In our choice of friends and associates, we are less constrained by accidents of geographical proximity. In our comings and goings, we depend less on the concurrence of others. We have more capacity to gain observational experience of an extended immediate environment. And for all of the preceding options, access is far more open and democratic than it was in preautomobile eras. Arguably, only the printing press (and perhaps within a few more years the microchip) rivals the automobile as an autonomy enhancing contrivance of technology.[16]

There are also important senses in which the control enabled by automobility has challenged hierarchies and constrained discrimination based on gender, race, and class. "Despite the violence and intimidation directed toward black drivers," Cotton Seiler observes of the United States in the first half of the twentieth century, "the road . . . to some degree provided a space where the everyday discrimination and coercion African Americans faced in other public spaces—in stores, theaters, public buildings, and restaurants, for example or on sidewalks and public transportation—could be blunted, circumvented, and even avenged."[17] This became more salient, he argues, with the rise of anonymous, placeless, limited-access interstate highways beginning in the later 1950s.[18]

Car ownership and the freedom to drive have also often been viewed as empowering for women. The movement to legalize women driving in Saudi Arabia indicates the attractions of such freedom in places where it does not exist.[19] More generally, robust cross-national evidence demonstrates that registration and ownership of cars is predominantly male (e.g., 64 percent in the United States; 75 percent in Sweden)[20] and that men drive considerably more—and for different purposes—than women.[21] Many conclude that this inequity is a constraint to be overcome, because, as one feminist scholar put it, "mobility and control over mobility both reflect and reinforce power."[22]

Expressed Preference in the Marketplace

Industry analysts have long referred to a "saturation point" for motor vehicle sales. Previously defined as one vehicle per household, it is now defined as a point at which every driver has access to one. By that definition, the United States today is oversaturated: there are *more* cars than there are licensed drivers.[23] Moreover, only about two percent of passenger trips use public transit in the United States.[24] Again, the United States is at the end of the spectrum in this regard, but the same trend can be found in a diverse array of societies. In European countries with both public policy and societal characteristics that favor public transit and bicycles over cars—costly fuel, excellent transit systems and bicycle infrastructure, and high population density—cars nonetheless now account for 80 percent of travel.[25] China's dramatic growth rate over the past two decades has reached 120 million passenger cars in 2013 (and 240 million vehicles overall) and is projected to add an additional 100 million over each of the next several decades.[26] Such projections are often presented as though they are describing a spontaneous natural phenomenon. China's history of state control over the economy and its continued constraints on political freedom make the recent explosion of driving and cars appear as the eruption of pent-up natural desire in spite of these limits. All this seems to reinforce the assertion of economist Charles Lave over two decades ago that "people increasingly and relentlessly choose the automobile over other forms of transportation."[27]

The sense that automobility is a reflection of consumer demand manifests another contemporary association of cars and freedom. The global growth of car-centric transportation systems is seen here as a response to the free choices of sovereign consumers in the marketplace. Such choice extends not only to vehicles themselves but to the lower-density, sprawling landscape of development that complements driving and is therefore an integral component of automobility.

A consequence of this naturalistic perspective on the global growth of automobility is that the search for structural alternatives to this growth, in the form of public transit and higher-density infill development, can readily be presented as paternalist or elitist. As a satirist, libertarian P. J. O'Rourke overstates the case, but it nonetheless remains salient to many: "Why do politicians love trains? Because they can tell where the tracks go. They know where everybody's going. It's all about control. It is all about power . . . Politicians hate cars . . . because cars make people free."[28]

Human Flourishing
The desire for more mobility is human nature.
—Daniel Sperling and Deborah Gordon[29]

The vast majority of trips taken with a car were not taken prior to its availability and could not be taken without it today.[30] This is a crucial point, as it defies arguments that better public transit infrastructure could, by itself, dramatically reverse automobility. Yet those added trips have enabled greater choice regarding where one lives and where one works, which need not be in close proximity either to each other or to a transit line. They allow for easier exit in search of better schools for one's children and they help accommodate the scheduling challenges of households with two wage-earning adults. They enable travel with less advanced planning and enable vacationing in more diverse and less congested locations. In all these senses, increased mobility appears to increase individual freedom. It is not merely the absence of restriction, or freedom *of* movement (a potential that need not be actualized; sometimes termed "motility"[31]) that is valued and sought to be maximized here but freedom *as* movement (mobility itself). It is the new trips that automobility enables us to take that are often regarded as increasing opportunities for human flourishing.[32]

Whereas the earlier manifestations of "autofreedom" discussed here focused upon the instrumental value of automobility to individual freedom, if it can be persuasively linked to human flourishing itself, then the increased mobility that automobility enables could be argued to have intrinsic value. Lomasky argues that "automobile transport is a good for people in virtue of its intrinsic features. Automobility has value because it extends the scope and magnitude of self direction."[33]

On the one hand, then, the argument for linking automobility to human flourishing itself is the most ambitious of the arguments outlined in this section and has the potential to encompass all of them. On the other, it most clearly brings to the fore those core questions about the nature of the good life. To the extent that we wish to engage critically the notion of autofreedom sketched to this point in the chapter, it is ultimately the argument that allows the greatest leverage.

Challenges of Automobility

Even ardent defenders of automobility concede that there are challenges. As James Dunn acknowledges: "The automobile is the solution to most Americans' transportation needs. But its very success has generated serious problems—most notably, congestion, pollution, and energy inefficiency—that need to be addressed by public policy."[34] In Dunn's formulation, these problems are negative externalities that can and should be mitigated without fundamentally altering the system of automobility itself. Yet the challenges are so extensive and intensive that it must remain an open question, at this point, how they might be addressed. Here my aim is simply to catalog a number of these challenges in order that we can subsequently explore approaches to change.

The problems acknowledged by Dunn certainly are among the most discussed and visible. The amount of time people spend stuck in traffic has become dramatically worse in recent decades in the United States and worldwide. Although there has been a plateau in the past several years in some places, the overall picture is of more people in more places stuck in traffic for longer times.[35] Even though emissions from new cars have been reduced dramatically over the past several decades, urban air pollution from cars also remains a serious problem due to the increase in their numbers and in vehicle miles traveled.[36] Finally, automobiles directly constitute about 20 percent of global emissions of carbon dioxide and other climate change gases—and considerably more in the United States.[37]

In addition to these high-profile problems associated with cars, we must add others: the relative immobility and dependence of non-drivers, especially children[38] and a growing elderly population;[39] the increasing tax burden in many countries to support an aging and sprawling automobility infrastructure; rising obesity correlated with car-dependent communities;[40] the tremendous percentage of land devoted to car use (two-thirds in Los Angeles);[41] the growing number of injuries and deaths worldwide, despite dramatic improvements in vehicle safety;[42] and geopolitical tensions, conflicts, and warfare caused or exacerbated by contested access to oil. One could readily add to this list.[43] Yet it does reinforce the point made by Leo Marx in this chapter's epigraph that we are well beyond the point at which we can meaningfully speak of the "impact" of the car on society; these problems and others constitute fabric of our lives and communities and so the horizon within which political theorizing and strategies for change must take place.

Many who discuss the practical challenges of automobility characterize the strategies to address them in a dichotomous manner. Two especially prevalent dichotomies frame the potential strategies as either technical or political and as either private or public.

Changes to Cars vs. Changes to Us

Strategies for change are typically presented as either technical, focused on changes to the car itself (e.g., hybrid or electric vehicles; improved design for fuel efficiency, safety, emission control, and recyclability; etc.) or they are political, focused on behavioral changes (e.g., promoting public transit and bicycling, carpooling, etc.) and structural changes to create new options and remake our communities in ways that reduce reliance on cars.[44]

This dichotomy—between changes to cars and changes to us—is often appealing to those who seek to defend automobility, because it seemingly offers one pathway to address problems through a technical fix—by changing cars—which appears to be consistent with autofreedom and avoids the more contentious sorts of strategies—and the complex inter-dependencies inherent in automobility—that appear to threaten this. Thus in 1954 California's governor could assert that "smog is a scientific and engineering problem and not a political or legal one."[45] Much more recently, the editor of *Road and Track* writes:

There is no denying that the automobile has a social cost—clean air, use of resources, accidents. In all fairness, these costs must be weighed against the benefits—mobility, freedom and independence. Manufacturers have done much to minimize the car's impact on the environment, energy and safety. Yet, despite these gains, there are those who can't abide these freedoms. . . . These arguments against automobility are cloaked in language about cleaning up the environment or improving fuel economy. But I believe it's convenient cover for a larger agenda that would dispense with the widespread use of the automobile—or at least create an environment where their use is severely curtailed or strictly controlled. . . . So, in some quarters, the view of the automobile and the freedom it confers on the masses have shifted from that of a social good to a necessary evil.[46]

Private vs. Public

The decades of explosive global growth of car ownership and miles traveled, in the face of many practical challenges such as climate change, congestion, and so forth, is often taken as a classic case of conflict between private desires and the public interest.[47]

This is true among advocates on both sides of this perceived dichotomy. Thus James Q. Wilson argued for privileging the private, stating that "the debate between car defenders and car haters is a debate between private benefits and public goods"; he concludes that this is "no real debate at all" given "the central fact that people have found cars to be the best means for getting about."[48] André Gorz, firmly positioned on the "car hater" side of this debate, also framed his argument in terms of this public–private divide, simply reversing the polarity and favoring the public side:

The worst thing about cars is that they are like castles or villas by the sea: luxury goods invented for the exclusive pleasure of a very rich minority, and which in conception and nature were never intended for the people. Unlike the vacuum cleaner, the radio, or the bicycle, which retain their use value when everyone has one, the car, like a villa by the sea, is only desirable and useful insofar as the masses don't have one . . . For when everyone claims the right to drive . . . everything comes to a halt, and the speed of city traffic plummets.[49]

Inside Criticism of "Autofreedom"

These dichotomies, between changing cars and changing us, as well as between private freedom and the public interest, can be useful to a point, yet they threaten to obscure more than they illuminate about the challenges of promoting freedom in a society defined by automobility.

The prospect of a technical fix to cars—the effort to "minimize the car's impact on the environment, energy and safety," in the words of the *Road and Track* editorial—holds the allure of avoiding threats to individual freedom. Conversely, structural and behavioral changes, including restrictions on driving, appear to be inherently at odds with autofreedom.[50] Yet neither side of this dichotomy is as clear as it seems.

Technical "fixes" have included new safety requirements and equipment, including shoulder belts, airbags, and requirements for crashworthiness. They have also included requirements for catalytic converters to minimize pollutants, outlawing of leaded gasoline, and fuel-efficiency standards. Vehicles that use hybrid or electric motors or other new fuels and new materials have also been categorized in this way. Of course, many of these "technical" changes have also been highly political—both in the sense that they entail active citizen engagement and contentious public debate about social ends and in the sense that they result in government mandates and regulations that constrain the power of auto manufacturers.[51]

Nor have these "technical fixes" escaped the concern for autofreedom. In addition to strong industry opposition, these changes have, at times, been resisted by consumers in part on the grounds that they would increase costs. Some—such as emissions testing requirements—also impose direct obligations on drivers. James Q. Wilson, who, we have seen, positioned himself as a strong proponent of cars, proposed and celebrated another sort of fix: cameras that monitor speeding and devices that could be used for measuring "the pollution of cars as they move on the highways and then ticketing the offenders."[52] Whatever one concludes about such surveillance-based approaches, they clearly cannot be counterpoised to a political approach and are hardly uncontroversial from the perspective of individual freedom. Equally evident is that these approaches resist the dichotomization of public and private. Chella Rajan characterizes this as the "enigma of automobility" as follows: "Cars serve to create privatized space for individual drivers, but driving propagates socially shared effects that could quite conceivably undermine the individualist credo of personal vehicle use."[53]

Or, as Mathew Paterson summarizes the point, these approaches to the challenges of automobility "start with simple technological devices but increasingly entail surveillance techniques and end up with wholesale management down to the level of individual journeys, driving techniques, practices, and so on."[54]

At the same time, although some changes have proven very successful at "fixing" the particular problem they were developed to address, others have been overwhelmed by the growth in overall number of vehicles and miles driven per vehicle. The result has been to mitigate what would have been an even greater problem but not to reduce it.[55] Because these strategies are targeted narrowly, they do little to modify other problems. Moreover, a number of challenges are generated by the sheer volume of traffic (land use, congestion) and these—by their very nature—are unlikely to be addressed through changes to the car itself. The technical fix can appear plausible—and "merely" technical—only by abstracting the car from the system of automobility within which it exists. Once that immense assemblage with its manifestations of power is recognized, the complex interdependencies of technical, behavioral, and structural changes becomes more evident and the challenge less difficult to perceive.

The very terms "public" and "private" also seem to be undermined by an automobility landscape. Vast amounts of space are devoted to car-only environments.[56] Although these are predominantly publicly funded and maintained roadways and even privately owned spaces such as parking lots and garages that guarantee public access, they challenge familiar notions of a public realm in the sense that they actively exclude or threaten nondrivers and anyone not in a vehicle. Conversely, drivers are in privately owned and operated cars, yet their licensing and behavior is heavily monitored and regulated.[57]

In sum, changes to the car appear practically inadequate to the challenges that society faces from automobility. Moreover, the very concept of limiting our strategies for change to those that are consistent with private interest and with changes to the technologies of the vehicle itself is based upon false dichotomies that obscure more than they illuminate. And yet, as noted at the outset, we cannot simply dismiss autofreedom as false consciousness. Such a move simply feeds the elitist caricature of automobility's proponents.[58] Thus what is needed is an inside critique of the components of autofreedom outlined in the first section of this chapter.

Revisiting Identity

I termed the first sense of autofreedom described earlier "identity" and sought to capture several ways in which being a driver has often facilitated a sense of independence and established a measure of adult participation in society. The car itself has also created a palette for individual and cultural expression.

And yet, the state apparatus used to train, evaluate, license, and monitor new and existing drivers is one element that weighs on the other side of this scale. Because they are operating a potentially deadly piece of machinery, automobility leads to a society that also criminalizes the risk-taking and other potentially foolish or experimental activities that are frequent rites of passage and can also reflect inherent developmental limitations of teens. Consuming alcohol and drugs are only the most obvious such activities; showing off for peers, misjudging one's level of attention or alertness, responding to a dare, and being seduced by speed are others. Freedom for such actions is much easier to tolerate in places where cars are not central, yet automobility raises the stakes so substantially that spaces for such freedom are far more circumscribed and "zero tolerance" is often the norm.

Structural and peer pressure to own a car also place important constraints upon independence; again this is reflected among many high-school-aged students and young adults who seek out paid employment—and substantial debt, with the many constraints this imposes—as a means of securing a car. Ivan Illich long ago captured this dilemma well:

The model American male devotes more than 1,600 hours a year to his car. He sits in it while it goes and while it stands idling. He parks it and searches for it. He earns the money to put down on it and to meet the monthly installments. He works to pay for gasoline, tolls, insurance, taxes, and tickets. He spends four of his sixteen waking hours on the road or gathering his resources for it. And this figure does not take into account the time consumed by other activities dictated by transport: time spent in hospitals, traffic courts, and garages; time spent watching automobile commercials or attending consumer education meetings to improve the quality of the next buy. The model American puts in 1,600 hours to get 7,500 miles: less than five miles per hour. In countries deprived of a transportation industry, people manage to do the same, walking wherever they want to go, and they allocate only 3 to 8 per cent of their society's time budget to traffic instead of 28 per cent. What distinguishes the traffic in rich countries from the traffic in poor countries is not more mileage per hour of life-time for the majority, but more hours of compulsory consumption of high doses of energy, packaged and unequally distributed by the transportation industry.[59]

Perhaps Illich's message is beginning to resonate more widely. In a recent cross-national comparison of eight postindustrial countries, the authors describe a phenomenon they label "peak travel": overall "travel activity has reached a plateau" and "private vehicle use . . . has declined in recent years in most of the eight," with notable declines in the United States and especially among young drivers.[60] This is consistent with some evidence that driving is becoming less integral to late adolescent and young adult identity than it has long been: even rates of possession of a driver's license among this group have dropped to its lowest level in half a century in the United States.[61] This is not simply a decline in "car culture," if that is understood simply as a spontaneous shift in values or lifestyle preferences.[62] In some measure, it reflects the availability of car- and ride-sharing programs and apps, and the rise of social media, smartphones and mobile electronic devices more generally.[63] It also reflects a generation coming of age during a period of persistent high gasoline prices and economic hard times—especially the declining economic fortunes of their own so-called millennial generation—and high levels of traffic congestion. Finally, in a number of places, it reflects the renewed attention to infrastructure for walkable, bikeable, and transit-oriented communities.[64] Although analysts struggle to tease apart the influence and implications of these factors and others, what is clear and significant is that this represents an opening and an opportunity for both a new sort of conversation about cars and freedom and for support and reinforcement of the sorts of structural changes that make reduced autodependence more feasible and more appealing.

Revisiting Control

Looking beyond identity, although the car can enable a flexible form of mobility over which the driver has substantial control these recent shifts indicate that it is not alone in offering flexible mobility. Perhaps more evident is that the control it promises in theory is often lacking in practice. The geographic separations of home, work, shopping, family, and entertainment often foster a dependence on the car and a frequent inability to avoid lengthy commutes, resource depletion, and the many other concomitants of living in communities structured around automobility. A simple example: when driving children to school becomes commonplace, a barrier to exiting from neighborhood public schools and seeking out others is radically reduced. The result, in my own community, is that even if one's children go to a neighborhood school, many of their classmates and friends are from other communities. Playdates, birthday parties, and other events and gatherings thus require automobile transportation even when attending school itself does not. As John Urry puts it, automobility "coerces people into an intense flexibility."[65]

Of course, it is the carless whose freedom is most evidently diminished in communities whose spatial geography is structured to the scale of traffic. Such geography results in greater dependence by children, the growing population of elderly, and others unable to drive, as well as those who do not own or have regular access to a car. Although drivers are frequently dependent upon their cars in these places, nondrivers are largely dependent upon drivers and non–car owners upon owners.

Recognizing these forms of dependence allows us to imagine alternatives as a strategy for increasing independence and control. Reducing both forms of dependence can be a result of fostering viable alternative forms *of* mobility, as well as alternatives *to* mobility.

This more nuanced understanding of the relationship of freedom-as-control to automobility invites us to revisit the persistent gendered differences in ownership and driving practices noted earlier. Recognizing these differences, it becomes clear that the notion that an automobility society enables control is one modeled to a greater degree upon the practices of men than of women. Yet what to make of the consistently reported fact that women car owners not only drive considerably fewer miles (roughly 25 percent fewer in the United States) and over a smaller spatial range but also drive for different purposes than men? That is, in the United States and many other countries women drive less often, and considerably shorter distances, to work. They drive more miles and take more frequent trips, however, for purposes related to social reproduction and care giving (e.g., grocery shopping, transporting children and elderly, household errands, etc.).[66] On one hand, consistent with the auto-freedom-as-control view sketched earlier, these differences in mobility might reflect a constraint—resulting in less adequate job opportunities, for instance. On the other hand, these differences might reflect a choice—one with much promise for envisioning more sustainable patterns of mobility and one which, again, can be supported and expanded through a wide variety of measures.[67] Where such measures are understood to facilitate and expand freedom—protecting freedom of movement (motility) even as they reduce freedom as movement (mobility)—they also become more broadly resonant and might then become more politically viable.

Revisiting Market Preferences

In densely populated urban areas developed prior to the entrenchment of automobility, the car is a liability for day-to-day needs, yet its frequent use makes the city itself less pleasant to live in than it would be otherwise and has often led these areas to be remade in an effort to accommodate traffic. Conversely, cars become more functional and necessary in lower-density, sprawling forms of development. Automobility is at the core of community design in these places, which devote a massive percentage of land to uses restricted to drivers, cars, and other related functions while being unavailable and inaccessible to all others. Such sprawl is thus a product of automobility but also a path-dependent prompt to its continuation.[68] What results is a transportation "monoculture" that dramatically constricts choice.[69] The market preference expressed for cars over other forms of mobility can only be understood in this path-dependent context and not as a reflection of some sort of autonomous decision making.[70] In a country like the United States where such a transportation monoculture is the norm, what is striking about the recent decline in vehicle miles traveled per capita is how much change there has been despite this. A truer test for the market preference claims for cars would require a much greater proliferation of both alternate forms of mobility and attractive options for the pursuit of livelihood, fulfillment, and pleasure—social reproduction—within a far more compact area.

Automobility and Human Flourishing

Finally, and most broadly, we must struggle to delink freedom from mobility in the sense that more mobility is necessarily understood to facilitate greater human flourishing. My freedom is clearly limited when my movement is restricted, but this does not mean that greater movement is an expression of greater freedom. Although I can express my freedom by looking for a job at some distance from my home, feeling trapped in a long commute can instead seem a source of unfreedom. The potential freedom expressed in the first instance is an idealized form that comes without the practical limitations, such as the commute, in the same sense that Jake's idea of the freedom of having a car came, at the time, without the actuality of monetary and work obligations that would finance it. This is not the only sort of obligation entailed by automobility. The opportunities for movement afforded by the car also "entail obligations . . . to be present in a variety of family, work and leisure time events and situations," obligations that are not always welcome.[71]

Not all mobility is desirable. When we are stuck in rush-hour traffic, when we have to take time off of work to transport an elderly parent to the doctor, when we shuttle our children to and from school, friends, or sports, this should be clear. The reason it is rarely recognized as such is because dependence on cars often comes to appear inevitable; the practice is embedded in the very structure and organization of our communities. When political choices lead to a different communal structure, one that facilitates proximity as much or more than movement over vast spaces, then the constraints imposed and the opportunities enabled for our freedom by the present system of automobility will become more evident. These choices not only make it more convenient to walk, bike, and use public transportation, they can make it more feasible and appealing to choose *not* to take a trip at all.[72] Only in this way can we achieve a level playing field upon which a real politics of mobility might emerge.

The key here is to recognize that this emergence is not something that we should look for in isolation from material conditions. Of the recent studies and articles noted previously that identify a sharp decline in driving among younger adults, many ask whether this change is a result of cultural change and choice or economic necessity or other material constraint. The suggestion is that in the latter case the change will be ephemeral. One recent article asks, "Are these the early adopters of an anti-automotive sentiment that soon will sweep the nation?" No, this author argues, because most who are carless have relatively low incomes. Hence, like others, he concludes that they are "carless by economic necessity rather than by choice."[73] This dichotomy is precisely the wrong way to frame the question. It presumes just what this chapter has challenged—the notion that driving and car ownership reflect, in this formulation, a universally "*pro*-automotive sentiment." Moreover, it presumes that lasting change depends upon postmaterialist attitudes and values, ones disconnected from economic or structural conditions. Viewed through a different lens, the recent decline in driving is a shift in practice to be enabled and built upon in ways large and small. The economic context of this shift in practice means that this support will improve the lives—and expand the freedoms—of many who are struggling economically.

This chapter represents an attempt to articulate a critique of what I have termed "autofreedom." In doing so, my goal has been to address a "hard" case. Rather than seeking to draw upon other values *as a trump to* the freedoms enabled by automobility, I have sought balance from within this widely acknowledged value. Doing otherwise raises the specter of paternalism, with its incumbent strategic and practical risks. Yet although such a critique is necessary, even in a more fully developed form it is unlikely to be sufficient. That is because there is a plurality of values to consider in relation to automobility—not just freedom. If we can cultivate a more balanced appreciation of the relationship between automobility and freedom, then we might enable a greater openness to this plurality. In that context, deliberation on human flourishing in relation to mobility can become more explicit rather than hidden behind an assumption that questions about automobility entail restrictions on freedom and—by extension—flourishing itself.

7

Homes, Household Practices, and the Domain(s) of Citizenship

If the core problem of wilderness is that it distances us too much from the very things it teaches us to value, then the question we must ask is what it can tell us about home, the place where we actually live. How can we take the positive values we associate with wilderness and bring them closer to home? . . . we need to discover a common middle ground in which all of these things, from the city to the wilderness, can somehow be encompassed in the word "home." Home, after all, is the place where finally we make our living. It is the place for which we take responsibility, the place we try to sustain so we can pass on what is best in it (and in ourselves) to our children.

—William Cronon, "The Trouble with Wilderness"[1]

The point is that all green actions in the home have a public impact, in the specific sense of the creation of an ecological footprint. This, in turn, potentially generates the kinds of obligations I have said we should associate with ecological citizenship.

—Andrew Dobson, *Citizenship and the Environment*[2]

We become fighters when something threatens our home.

—Cora Tucker, Environmental Justice leader[3]

As this chapter's epigraphs indicate, "home" has become a normatively appealing foundation for environmental arguments in the past couple of decades. Indeed, as many have noted, the etymology of "ecology" connects it with the ancient Greek "*oikos*," the home or household.[4] Yet as the context of each of the quotations suggests, home also stands in contrast to other, more conventional ways of conceptualizing ecological concern and action. For Cronon, it is a contrast to the "trouble" with "wilderness," conceptualized as a place distant—both conceptually and spatially—from where we live. For Dobson, arguing that the obligations of citizenship begin at home is at odds with dominant conceptions of citizenship as action circumscribed within the public sphere. For Tucker, organizing motivated by threats to home—where we "live, work, and play"—set the environmental justice movement apart, in both style and substance, from other manifestations of environmental activism.

I applaud this emergent recognition of home and household as an integral concern that environmental critics must engage, because this is perhaps the premier location of material practices central to everyday life. In postindustrial societies, one study has recently noted the following:

Households serve as a reservoir into which resources are sunk, as a center where purchasing decisions are made, as a nexus for energy used in transportation and heating . . . The typical infrastructure associated with houses sucks in resources through roads, water lines, gas lines, power lines, internet cables, and phone lines. Homes also come equipped with the means to excrete wastes, through sewer systems and trash-collection networks. This system makes households the nexus for resource consumption.[5]

The number of such homes is also on the rise—independent of population growth—due to declines in the average number of people per household.[6]

The centrality of home and household to environmental concern is reflected in other broad cultural trends. Indeed it seems fair to say, with Kersty Hobson, that "there is now a general tendency of individuals in countries such as the United Kingdom to demarcate the home as the main space where pro-environmental action takes place."[7] Yet identifying the home as "the main space" for "pro-environmental action" invites crucial questions about what sorts of action are included and excluded here, as well as questions about the broader implications of situating action here.

One of the most familiar responses to this newfound attention has been the rise of "ethical" or "green" consumerism. Here we find a diverse and growing array of products, services, and practices—as well as advice and examples on websites, blogs, magazines, and books—that has emerged in a proclaimed effort to make homes, houses, and household activities more environmentally responsible. No doubt familiar to most readers, examples include lower-impact building materials and design, solar panels and waste-reduction practices, recycled products, nontoxic household cleaners, energy-efficient appliances, local and organic foods, and reusable bags and bottles.

Many have rightly charged that such a consumerist approach threatens to narrow our imagination of feasible alternatives by privatizing and individualizing responsibility. Some also draw attention to the exclusivity and privilege reflected in an approach that often depends upon a combination of money, time, access, and sometimes specialized knowledge for action—a combination most likely possessed by those identified by Inglehart as "postmaterialists." Those seeking an alternative to this privatized green consumerism are often drawn to a notion of public citizenship instead. I argue, however, that this contrast between private consumerism and public citizenship overlooks much that might be promising in attention to home and household practices, a promise indicated by activists such as Cora Tucker who are motivated to become "fighters" by threats to—and concerns with—their home. As the nexus for a vast array of infrastructure, resources, and practices, home and household are integral to the reproduction of everyday life. Securing food, clothing, and shelter, caring, cooking, cleaning, making, provisioning, nurturing, teaching, welcoming, excluding, fighting, living, and dying—*homemaking*—is a material practice. As such it is quite simply a far broader terrain and crucible for action than the familiar characterization of "greening the home" through private consumer choices would lead us to believe. I argue that reflection upon homemaking can prompt us to think in fruitful new ways about the possibilities for environmental criticism and change. To do so, however, we must avoid the individualization of responsibility that often accompanies green consumerism, attending instead to collective and political action—a crucial element of citizenship.

I begin by reflecting upon a decidedly unconventional home. My experiences here challenged me to revisit some of my own preconceptions by highlighting tensions, questions, and possibilities that point toward the sort of citizenship that I develop later in the chapter.

An Interlude and Illustration: CCAT

Shortly after my arrival as an assistant professor at my university in the late 1990s, I was invited to give a presentation at a student-run organization affectionately known on campus as "see-cat"—the Campus Center for Appropriate Technology (CCAT). Although I had a previous background in student activism, CCAT stood out among the many student organizations with which I was familiar; for one thing, it was a rough-around-the-edges house with a constant hum of activity surrounding it. It was the residence for three students, selected annually to codirect its programs, and a home away from home to a great many other members of the campus community. Over the years, CCAT has developed and implemented a wide array of renewable energy technologies, energy- and water-conservation strategies, low-impact building techniques, materials reuse, and organic gardens. It has also cultivated lots of pragmatic, knowledgeable, and hardworking student activists and leaders.

Founded twenty years earlier, in the late 1970s, CCAT's name hearkens back to its emergence in the heyday of the so-called appropriate technology movement and was modeled after the "Integral Urban House" in Berkeley, California—an effort to apply some of the self-reliant "back to the land" ethos of the era in a more urban setting.[8] Appropriate technology as a whole was inspired by the economic and cultural analysis of works such as E. F. Schumacher's *Small Is Beautiful*.[9]

Critics have long contended that what the movement lacked, however, was an explicit conception of how their envisioned social model was to be achieved; that is, how "appropriate" forms of technology might become integral to society as a whole rather than an "alternative" attractive only to those on the margins. For example, Langdon Winner argued that the movement implicitly embraced what he labeled the "build a better mousetrap" approach to social change. This approach relied upon an unsubstantiated faith that the creation and development of new forms of renewable energy, smaller-scale technologies, and other practices could become widespread simply through the inherent power of a good idea and thereby bring about a social revolution. Because the advocates of appropriate technology were inattentive to or dismissive of the role of social and political institutions and power, Winner writes elegiacally that they were "lovely visionaries, naïve about the forces that confronted them."[10] With this context in mind, it is not surprising that the center of gravity at the time of CCAT's founding was among students learning to build better mousetraps while focusing relatively little on social and political power: those majoring in environmental engineering.

By the time of my arrival in the 1990s, the very language of "appropriate technology" as well as the "mousetrap" approach seemed to me anachronistic, of a piece with the aging hippies who could still be found strolling the plaza of our small Northern California town. Sympathetic as I was, being invited to talk about the future of appropriate technology was a challenge, because it wasn't immediately apparent to me that there was such a future, or what constructive role this organization and the ideas that inspired it could play in contemporary efforts on behalf of environmental sustainability. Nonetheless, I accepted the invitation in large part because of the infectious energy, conviction, and talent that emanated from the students involved at the house. It was evident that although the language of "appropriate technology" had faded from the wider public discourse the appeal of CCAT's vision was strong. Wrestling with these tensions has been central to my involvement with CCAT over the years. Here I want to sketch some possibilities that emerge from these tensions, in a way that has broader implications for thinking about home, citizenship, and environmental criticism.

Shortly before writing this, CCAT celebrated its thirty-fifth anniversary, an impressive achievement for any group and a rather remarkable one for such an ambitious and student-run endeavor. Nonetheless, what Winner characterized as the technological exuberance and individualism of the mousetrap theory of social change has long resonated with me. I have come to conclude that one secret to CCAT's success is that their vision is no longer—if it ever really was—driven by the mousetrap theory.

Indeed, the vision and mission statement of CCAT is more reflective of its context as a home and household than the mousetrap theory would lead one to expect. Here CCAT is characterized as committed to demonstrating a way of living that is "both practical and rewarding." Language of community building, bringing together diverse viewpoints, and learning by doing are deeply entwined with the more expected references to sustainable technological systems themselves.[11] First and foremost, I have come to see, CCAT is an experiment in living, and its projects and activities are the embodiment of an ongoing dialogue about what sort of life it is good to live and what sort of household practices might help facilitate that life. The interests of the student directors is also suggestive here: although I have noted that the early leaders were drawn predominantly from engineering, a far higher number now come from majors in the humanities and social sciences, in which dialogue, critical thinking, and attention to questions of power and institutions are central to the endeavor.

To discuss home—CCAT or other—requires that we recognize the significance of two relevant aspects. Heidegger characterized these as "building" and "dwelling" while going on to argue that "to build is in itself already to dwell."[12] The first of these—building—results in the physical structure of shelter, or *housing*. The second—dwelling—entails a wide diversity of household practices. These two sides of home are integral and exist in dynamic relation to each other. To insist upon the centrality of both is important, because, as Iris Marion Young has shown, even Heidegger—who begins by positing this relation—goes on to privilege the work of building while neglecting the work entailed in dwelling.[13]

Young critiques Heidegger's attribution of creativity distinctly to building as a male bias, only conceivable by presuming that women occupy a fixed and essentialized role as those who cultivate and preserve the dwelling space once built. Stabilized in this way, it would appear that there is far less to say about the aspects of dwelling that aren't focused upon building itself. The rest can seem to be mere "housework"— the endless, repetitive, gendered, and instrumental labor of cleaning, cooking, and maintaining. Young exposes and rejects this presumption that "all homemaking is housework," however, thereby allowing us to recognize the cultivation and preservation of dwelling as active processes of engaged subjects, which can only exist in dynamic relation to building.[14]

The engineering students who founded CCAT were necessarily focused upon building—or, in fact, rebuilding. The house that they occupied and renovated was a vacant and dilapidated residential structure in a neglected corner of the university campus, slated to be set alight as training for firefighters. In relatively short order, and with the university's tacit permission, they garnered extensive donations of money and materials, renovated the interior, rewired the electricity, poured insulation into the walls and ceiling, installed active solar technology, and began a longer process of developing and installing rainwater catchment, gray-water treatment, a composting toilet, and a greenhouse that also provided passive solar heating, while also reclaiming the surrounding land for vegetable and herb gardens.[15]

Although hardly immune to gendered norms and stereotypes, as a residential student-run program rather than a nuclear family unit they could not presume that the household practices of cultivation and dwelling would be fixed based upon gender and made largely invisible. That's not to say that such matters always went smoothly. In fact, many interpersonal conflicts and negotiations, some of them gendered, centered upon such matters as cleaning, caring for the house's (very) public spaces, and other essential household practices. As the technologies and systems of the house have expanded, so have the diversity of practices integral to dwelling within it. Building—especially for a house like CCAT—is an ongoing endeavor. Nonetheless, for substantial periods in CCAT's history, these projects have receded in importance while the focus on dwelling became primary. Here the question is how to live with—and within—the building, a dynamic and contested process that can never be determined by the systems or technologies themselves. It is in relation to these household roles and responsibilities that it becomes clear that as an experiment in living CCAT is necessarily about dwelling in the broadest sense, not just building, and that practices of dwelling are inescapably political.

Although the 1970s language of appropriate technology has not made a widespread comeback, the notion of a more environmentally sustainable home, and attention to household practices as central to sustainability, certainly has. Moreover, the notion of a model or demonstration house has become widespread in recent years. It's not just LEED-certified or other so-called sustainable buildings and systems that have garnered attention; public accounts—often in the form of blogs, books, and other narratives—of a household's practices have also become widespread in recent years.[16] CCAT anticipated much of this recent growth of model homes and sustainable living experiments by several decades. As Noortje Marres shows, the structures and systems of these homes, as well as technologies designed to monitor, document, and display their energy use and other sustainability-related features, invest these things with normative meaning that make them a public basis for political contestation.[17]

I've served as a member of CCAT's advisory committee for many years now. Composed of present and former student leaders as well as other university and community representatives, this group has provided one of several spaces for regular discussion, debate, and evaluation of current and proposed projects, plans, and systems at CCAT. Meetings provide an opportunity for the more technically knowledgeable members to contribute their expertise in evaluating rapidly changing technologies and building techniques. Yet far more fundamental is an ongoing conversation about the ends and goals the group ought to be pursuing and the particular interpretations and values that members bring to bear on the inescapable and never-finally-settled questions of what counts as "sustainable" or "appropriate"—not in the abstract but in the concrete, lived reality of CCAT's residents and many other participants and visitors. In this sense, the material practices of the household and systems of the house are best understood in Latour's terms as conversations about "matters of concern" rather than fixed, uncontested "matters of fact."[18]

I'll offer two illustrations of this conversation here. It was a point of pride and publicity years ago when the group cut the cables tying the house to the electric utility's power grid. CCAT had assembled an extensive rooftop solar panel array early in their history and maintained a basement battery bank that (when properly managed) supplied the house with a steady flow of electricity, even in our foggy coastal climate. Indeed, during statewide energy shortages and rolling blackouts that struck California in 2001 this independence from the grid became a source of substantial media interest, generating statewide and national news coverage.[19] Yet not long afterward some were arguing that a newly feasible "grid intertie," whereby the rooftop solar energy system allowed power to flow back into the electrical grid while the house simultaneously drew its electricity from the grid, was a more "appropriate" system. In part, this discussion revolved around the comparative technical strengths of battery systems and grid reliance. The assessment also entailed an evaluation of the embedded energy and toxicity of the battery bank. Even more integral to these discussions was the question of just what sort of values these differing systems reflected. Among the most persuasive arguments to many was one that associated the batteries with *in*dependence and the grid with *inter*dependence. Framed in this manner, what had been a divided group coalesced around the "appropriateness" of acknowledging and modeling interdependence.

In roughly this same period, the organization confronted one of the most existential questions in its history. The house that was CCAT was within the planned footprint of a large new academic building slated for construction, a development the organization was powerless to prevent. As part of this building project, university officials proposed to move the organization to a house to be newly, and professionally, built for this purpose. After extensive consultation, discussion, and debate—and to the surprise of many—the organization turned down the offer of new construction. Instead, they chose to move the aging, existing house—with all its quirks, many student renovations, and extensive history—onto a new and more expansive foundation a short distance away. For those outside the organization (and some on the inside) who regarded CCAT first and foremost as a demonstration site for cutting-edge sustainable technologies, this decision was hard to fathom.

Yet, as should already be clear, the organization's primary identity was never rooted in this model. Instead, students past and present saw in the existing house *their* house, which reflected a deep attachment as well as a demonstration of their success in creating a sustainable household by building upon and working with what already exists rather than starting from scratch. Living in a home embedded with the labor and creativity of previous generations of students was valued not only by those from the past but also by the then-current generation of students. Here the decision reflects the understanding of home articulated by cultural geographers such as Alison Blunt and Robyn Dowling, who argue that "home . . . is a *place*, a site in which we live. But, more than this, home is also an idea and an imaginary that is imbued with feelings" and that ultimately it is "neither the dwelling nor the feeling, but the relation between the two."[20] Moreover, by working to improve what they had rather than building from scratch, CCAT actively chose to model the challenges involved in promoting sustainability in existing homes and communities rather than modeling the technological possibilities available when designing upon a blank slate.

"Sure," one might observe, "CCAT might be a home, of sorts, but it is hardly typical of the dynamics or challenges of many other homes and households." A reasonable point, but this doesn't cut as deeply as it might initially seem. First, like CCAT, many households are composed

of unrelated members: 2010 census statistics in the United States identified nearly 34 percent as "nonfamily households," including people
living alone or nonrelatives living together.[21] Second, like many other
households, CCAT members return frequently, maintain ongoing connections, and remain invested in the preservation of its past, the vitality
of its present, and promise of its future. In the end, it is to a large degree
these familiar elements of home, more than the specific technologies
operationalized on the site, that make the organization lasting and
significant.

When Langdon Winner critiqued the 1970s appropriate technology
movement by describing the limits of its "mousetrap" theory of social
change, he was only focusing in part upon the movement's technological
orientation itself. More urgently, he worried that it reflected a retreat
into the private sphere and so an avoidance of difficult public challenges
of power and politics. Appropriate technologists "dropped out of political activity and began a certain kind of sociotechnical tinkering: roof
gardens, solar collectors, and windmills."[22] This location in the private
sphere led to an implicit faith in consumer choice—and the supposedly
invisible hand of the marketplace—as the driver of change. As he put it:
"People would, in effect, vote on the shape of the future through their
consumer/builder choices. This notion of social change provided the
underlying rationale for the amazing emphasis on do-it-yourself manuals,
catalogues, demonstration sites, information sharing, and "networking"
that characterized appropriate technology during its heyday."[23] CCAT is
not immune to this sort of notion. Yet an apolitical retreat into the
private is not consistent with the actual practice of the organization or
those drawn to it, which intimates a different approach to change. Developing these contrasting approaches here will prove useful.

I have already suggested why this organization and its home cannot
be properly understood as a mere display of technologies. Such a display
would be largely static (or would be upgraded at set intervals to incorporate advances in technology), and its builders would be largely invisible or irrelevant to its purpose; hence the question of whether students
or professionals did the work would be unimportant, apart from matters
of overall quality and aesthetics. In such a display home, visitors would
be cast primarily as prospective consumers, exposed to staged demon-
strations of gadgets and systems that they might later purchase or adopt

for their own home. Interestingly, such a model would also encourage invisibility for the very technologies being demonstrated and the material flows that they regulate, just as these are invisible in conventional homes in postindustrial societies. The demonstration is most effective if solar electricity powers lights at the flip of a switch just as in "regular" houses. The kitchen sink could be connected to a gray-water marsh, but it would still drain water in the "usual" way. Here, green or appropriate technologies reduce resource usage for a given task but are regarded as fixed goods to be consumed in a manner that minimizes visibility and impact upon household practices.

By contrast, I have argued here that CCAT is better understood as an experiment in living and an ongoing conversation. The "appropriateness" of the technologies and systems are not presented as fixed properties of off-the-shelf products or designs but are tentative assessments of the particular fit between particular strategies or systems, the character, location, and practices of those using the house itself, and the values embraced by the current generation of student leaders. Such assessments require that systems themselves be visible—in order to be monitored, discussed, and evaluated. These elements can be encompassed in a form of democratic citizenship, in which the conversation is always embedded in the experiment; that is, the material practices are always at the core of the reflective discussions, judgments, and decisions made.

Conversation and action emanate from student residents' intimate participation in the complex assemblage of material flows and practices that encompass the household and home—including its surrounding gardens and other facilities. The spaces within and around the house are a hub for countless hands-on workshops, meetings, speakers, and other educational events, as well as volunteer workdays, concerts, and other performances. The line between "private" living spaces and "public" gathering spaces is subject to ongoing negotiation and change, but it is evident that both physically and metaphorically that line is fuzzy. Moreover, those who make the house their home do not simply maintain the house and program nor even lead the organization.

The experience both of living with such a rich array of technologies, gardens, and other systems and of negotiating their preservation and development with fellow students is transformative for virtually all of them. With real-time monitors of their energy consumption, they become much more aware and knowledgeable about the sources and magnitudes of impact by various activities of daily life. By living comfortably with systems that allow them to minimize their ecological footprint, they are able to challenge narratives that equate such living with deprivation. Yet—perhaps surprisingly—by living amid technologies that remain outside the mainstream experience of many, they also become acutely aware that these are necessarily entwined with decisions and priorities about how to live well—a broad and apt definition of politics itself. The consequence of these experiences is that they not only act as organizational leaders but become far more grounded, knowledgeable, and engaged citizens in the course of doing so.

As noted previously, Sherilyn MacGregor argues persuasively that much of women's environmental activism is best conceptualized in the public language of citizenship rather than the private language of care. MacGregor argues that although the private roles of mother and caregiver are often deployed strategically by activists in this context their engagement is rarely a straightforward reflection of these roles. Like the CCAT students, typically those drawn to the cause are transformed as they become active, public leaders. As Robert Gottlieb described: "Engagement in the movement . . . has often transformed how [these women] perceive their own identities as well. As organizers and leaders who become capable of questioning and challenging various sources of institutional power, many of the women in the antitoxics groups are transformed, in the eyes of others, into different people."[24] The fit between CCAT's decidedly public character and both MacGregor and Gottlieb's accounts is a good one. Those who have lived in the house have gone on to become planning directors and teachers, activists, community-supported farmers, NGO leaders, and policymakers. The home became a base for lives as active citizens rather than a retreat into private household practices or technological tinkering.

I don't wish to make too much of my reflection on the CCAT experience. Of course, it remains an atypical household in many respects, and the broader lessons that can be drawn from it are necessarily constrained by this reality. Yet just as sites like CCAT can help us imagine new ways to manage material flows of energy, water, waste, and resources in and out of the home, here I have argued that it can help us to see how the attentiveness to these flows might facilitate a form of citizenship that goes beyond private consumption. In this sense, once again, CCAT is suggestive of the never-purely-private character of home.

Rethinking the Relationship between Citizenship and Consumption

The practical connection between household and citizen action found at CCAT is, I argue, broadly relevant. Yet much existing thinking and writing about citizens and consumers hampers our ability to conceive of and develop this connection productively. On the one hand, much theorizing about citizenship in general and environmental citizenship in particular posits a dichotomy between this and a private realm of consumption and the home. On the other hand, by contrast, a fair amount of recent rhetoric seemingly conflates the role of household consumer with that of citizen. I analyze both approaches here in order to set the stage for the next section, in which I sketch an alternative that takes the distinctively political qualities of citizenship seriously while recognizing the ways these qualities can and have emerged from the material practices of the home.

Mark Sagoff has argued eloquently that most people think and act differently depending upon the role they conceive themselves playing. The result is that environmental challenges will be addressed very differently if we begin from the values we hold as public citizens than if we begin from the self-interested preferences we express as private consumers—either through our actual purchases in a marketplace or when queried by economists or pollsters about our willingness to pay for our preferences. Sagoff illustrates this claim, in one prominent example, by contrasting his university students' opposition to developing a picturesque valley into a grand ski resort with their apparent eagerness to patronize the resort if it were nonetheless to be built.[25] Rather than decrying his students' hypocrisy or the incoherence of their positions, he argues that their likely consumer preference—to ski the valley if a resort is available—is simply not a measure of the same thing as their citizen valuation of the undeveloped valley. This is commonplace, Sagoff argues: citizens "may themselves condemn the likely consequences of their own consumer interests on cultural or ethical grounds."[26] He goes further, personalizing the argument by asserting:

I love my car; I hate the bus. Yet I vote for candidates who promise to tax gasoline to pay for public transportation. I send my dues to the Sierra Club to protect areas in Alaska I shall never visit. And I support the work of the American League to Abolish Capital Punishment although, personally, I have nothing to gain one way or the other. (If I hang, I will hang myself.) And of course, I applaud the Endangered Species Act, although I have no earthly use for the Colorado squaw-fish or the Indiana bat. The political causes I support seem to have little or no basis in my interests as a consumer, because I take different points of view when I vote and when I shop. I have an 'Ecology Now' sticker on a car that drips oil everywhere it's parked.[27]

The point of this is (presumably!) not to encourage us to ignore oil leaks from our cars or—as Sagoff also claims to have done—to bribe a government official to fix traffic tickets.[28] It is, instead, to argue that laws and regulations that express our citizen values are both legitimate and important precisely because they don't reflect what we would otherwise choose as individual consumers. Written in the 1980s when memory of the passage of the sweeping environmental legislation of the early 1970s was still relatively fresh—yet under attack—such an approach offered a powerful justification for such laws. Today, when my own students read Sagoff's account, they—like me—still find his conceptual distinction between consumer and citizen roles illuminating. Yet at the same time many are both surprised and troubled by his account of the disparity in these expressed views—by his students and by him.

Two related changes may help explain their perception. First, the notion that far-reaching environmental legislation could be adopted despite explicitly conflicting with self-interested consumer preferences strikes many of my students as unlikely if not implausible. Why wouldn't it? Few if any examples of ambitious legislation of this sort have been adopted in their lifetime in the United States, and popular explanations for the more recent failures of legislation on climate change often explicitly point to opposition based upon consumer preferences.[29] Second, the idea that a concerned environmentalist would or even could ignore all of the manifold opportunities and consumer products designed to allow for the individual exercise of their values—from energy-efficient light bulbs and reusable bags to recycling, solar panels, and fair-trade or locally grown foods—seems to them to require willful avoidance. None of us are saints, my students declare, but the litany of personal behaviors that Sagoff presents strikes many of them not only as hypocritical but arrogant.

Writing more recently than Sagoff, Michael Maniates noticed the shift that I'm describing here. Maniates diagnoses and criticizes the evident growth of green consumer products and services in the 1990s and beyond as a reflection of an "individualization of responsibility" that both narrows the "environmental imagination" and displaces the possibilities for public, citizen action.[30] It narrows our imagination, Maniates argues, by excluding all those possibilities that require laws or collective action for their realization. Individual consumers might chose to purchase a hybrid car, for example, but they cannot create adequate bicycle or public-transit infrastructure through market choice alone. A singular focus on consumer choices displaces the difficult and uncertain work of citizen action to achieve such possibilities by offering the false promise of an easy way to "make a difference."[31] Despite these differences, the sense that public citizenship stands in contrast to private consumerism is shared by both authors; both also point to the promise of the former over the latter. Yet these shared perspectives have been called into question within the growing discussion of "green," "ecological," or "environmental" citizenship over the past decade.

As noted at the outset of this chapter, Andrew Dobson has sought to radically reconceptualize citizenship by bringing it "home," arguing that meaningful environmental action requires drawing the private sphere of the household into the ostensibly public sphere of citizenship. For Dobson, this "ecological citizenship" was an inclusive move that entails a radical departure from dominant liberal and republican conceptions of citizenship.[32] Although Dobson's incorporation of home within a conception of citizenship remains a promising direction that I wish to develop in this chapter, in the hands of many subsequent scholars and (especially) practitioners it appears to have morphed into a presumption that private individualism and green consumerism is a—perhaps *the*—primary manifestation of citizenship itself. Rather than publicizing private household practices, this appears as the privatization and individualization of the concept of citizenship. In this sense, environmental citizenship has come to "represent growing interest among scholars, policy practitioners and NGOs in the roles that *individuals* can, do and should play,"[33] and reflects a growing equivalence between citizenship and consumption. This has become so commonplace that Sherilyn MacGregor is able to conclude that "consumption . . . is now central to visions of environmental citizenship."[34] From the analytical perspective of Sagoff and Maniates, there is a through-the-looking-glass quality to this new equivalence, highlighted by another recent critic who is able to characterize contemporary notions of environmental *citizenship* as "individualising environmental responsibility"—precisely the language that Maniates had earlier used to critique green *consumption*.[35]

There is thus good reason to worry that attention to the role of homemaking exacerbates the tendency toward an individualization of responsibility. The limitations of microlevel, individualized consumer actions as leading to macrolevel, social change need to be taken very seriously. Yet the centrality of home to everyday life invites a more nuanced consideration of its potential to generate broader forms of collective or political action. This potential can only be realized if we can disentangle it from the dangers of individualizing responsibility. When home is conceived as a private, apolitical site for consumption, then these dangers are at their greatest. Green consumerism and behaviors appear as a lifestyle choice; an act of will that entails embracing or crafting a particular—and limiting—identity as an "environmentalist." By contrast, when home is understood more broadly—as the site central to the reproduction of everyday life, the nexus for a wide variety of material practices and flows—then the already present concerns and identities that emerge from the home can be recognized as a basis for environmental action.

Politicizing Material Practices of Home

Home as Haven

The dominant conception of home today—in the United States and many other postindustrial societies—approximates what Dolores Hayden has termed the "home as haven."[36] Envisioned in the nineteenth century as a self-contained private refuge for a nuclear family nurtured by the wife and mother, "the spatial envelope for all of this exclusive nurturing was a little cottage in a garden."[37] It is really since the mid-twentieth century that this conception—often manifest as the suburban house built as a part of a large-scale, sprawling development of similar houses in a community dependent upon automobility—has become fully manifest. Although this form of development is particularly associated with the

United States and Australia—where over half the population now lives in suburbs—manifestations can also be found throughout North America, Europe, and increasingly around the world.[38]

There have been many critics of the sprawling landscape of so-called greenfield development (that is, building houses on the farmland, fields, or forests at the periphery of existing cities and towns) in recent decades. Critics have charged suburban sprawl with fostering a sense of placelessness, aesthetic violence, and anomie.[39] They have also highlighted concerns about growing commuting times, lack of public facilities and services, destruction of agricultural land and wildlands, and dramatically expanded energy consumption and environmental impact.[40] These criticisms have also led to countermovements that emphasize designs that foster walkability and integration of business and residential uses, as well as in-fill and transit-oriented development. In the United States, these efforts are often captured under the labels "new urbanism" and "smart growth."[41] Although these efforts have gained a foothold in the United States and both reflect and have encouraged a renewed attraction to urban living, the juggernaut of low-density, monocultural, sprawling development has continued to expand its footprint.[42] In the years leading up to the 2008 economic crash, this juggernaut was fueled by subprime mortgages and other "creative" financing instruments. Even as home building has revived more recently, however, little that is fundamental seems to have changed.

A similar double movement characterizes the practices taking place within these homes. On the one hand, we can identify household practices that seek to reduce material flows and ecological impact. On the other, the cultural and economic preoccupation with home improvement and consumption—fueled by the growing sector of stores, products, services, and media cultivating an expanded conception of the ideal home— has also grown in recent years.[43]

Envisioning meaningful alternatives that would cut more deeply into this juggernaut first requires that we recognize some of the attractions— both for those who live in these homes and those who aspire to do so.[44] These can entail a sense of safety and security, privacy, and control, as well as the potential equity that might be facilitated by purchasing a private home as a consumer commodity.[45] Yet the closer we look at the practices of home and household, the less they fit the conception of an autonomous private haven that many imagine. Once we move beyond this conception, we can imagine new possibilities that draw together practices in the home and citizenship.

Citizenship and the Home: Three Aspects

Home is not a wholly private realm disconnected from the public. Both materially and conceptually, it is embedded in a web of connections and flows with the outside. I focus here on the way that our material experiences as home dwellers are an important subject of citizenship in general and of citizen action to promote sustainability in particular. We must recognize that the public and citizenship already has a place here. Advancing this claim requires being clear about its distance from familiar notions of green-lifestyle consumerism. These can be imbued with privilege, often appealing primarily to that so-called postmaterialist population that has the financial resources, time, and inclination to seek out these products and services.[46] Moreover, conceiving of environmental action in terms of household purchases shifts the onus to individuals—and most often to women.[47] Unlike those who equate environmental citizenship with consumption, I aim to reinforce the distinctiveness of citizenship in a society where it is already often devalued. There are at least three distinct senses in which citizenship emerges in relation to home.

First, familiar public notions of citizenship have a long history of connection with the home. Fiona Allon is quite correct that "home and home ownership in particular have long existed as powerful forces for individual and collective identity formation and self-definition, playing a crucial role in constructions of citizenship and national identity."[48] Normative weight has often been attributed to this point, asserting that stable home dwellers and especially owners possess the deep ties to the fate of the polity that make them good citizens. Historically, of course, claims of this sort have been the basis for excluding nonowners from rights of citizenship, such as voting. Yet the point is equally salient for critics, such as David Harvey, who characterize home ownership—and especially mortgage holding—as a means to pacify the citizenry: "The suburbanization of the United States was not merely a matter of new infrastructures . . . It also altered the political landscape, as subsidized home-ownership for the middle classes changed the focus of community action towards the defence of property values and individualized identities, turning the suburban vote towards conservative republicanism. Debt-encumbered homeowners, it was argued, were less likely to go on strike."[49]

With regard to practices within the household, feminist theorists have long and rightly insisted upon their integral connection to citizenship. The distribution of roles played by adults within the household have a substantial effect upon the ability of these adults to attend, participate in, and contribute to governmental meetings; to lodge complaints against public officials; to participate in community organizations or protests; and to become involved in other conventionally recognized forms of public discourse and citizenship. To the extent that women working outside the home are frequently assumed to be primarily responsible for a "second shift" as primary caregivers and homemakers, either because of a gendered division of responsibility or because they lead a single-parent household, their participation in these forms of citizenship will be—all other things being equal—more constrained than many men's. Moreover, the relationships and practices within the home are a vital crucible for the cultivation of future citizens: children.[50]

Second, there is a critical and often unrecognized sense in which the material environment of our homes and neighborhoods shapes, enables, and constrains participation in public life as citizens. Thad Williamson has drawn upon substantial empirical data to show how low-density sprawling communities—despite what he acknowledges as their considerable attractions and benefits for many—are "also constituent of a way of life that prioritizes privatism and consumerism over engaged political participation and ecological sustainability."[51] Higher-density housing, which facilitates closer access to services and transportation options, by contrast, can encourage a different set of priorities. In a complementary fashion, Dolores Hayden highlights the many ways in which the home designed as a haven has required a tremendous—yet neither inevitable nor unavoidable—amount of time devoted to household and caring practices, primarily by women. Only by recognizing this can we begin to envision alternative conceptions of home that result in a more widely shared responsibility for these practices. Far from being the singular or natural formation for everyday practices, including cooking, cleaning, and dependent-care, Hayden demonstrates that both this formation and alternatives have a particular history that can be traced back to the nineteenth century.

Alternatives that Hayden describes as embodying a "neighborhood strategy" are those in which household practices, as well as appliances for cooking, cleaning, and maintenance, can be shared collectively, thus lessening the time and financial burden on any single individual or household.[52] In the twenty-first century, this strategy finds expression in alternative models of home and neighborhood design, including cohousing projects, where private living space is conjoined with a kitchen and other facilities to be used collectively—for some shared meals, shared childcare, garden, or recreational facilities, and shared bedrooms available for guests.[53] Hayden, along with other architects, planners, and developers, has also envisioned practical ways in which the widespread, existing low-density suburban housing block can be (and has been) reimagined and restructured to meet the needs of a more diverse population by carving out existing, privately owned yard space for common activities such as child- or eldercare, laundry, food gardens, or shared kitchens, as well as options for small accessory apartments for single residents or couples—adult children, elders, or tenants.[54] It also finds expression in the so-called sharing economy—a rapidly growing array of (often Internet-mediated) networks for sharing bicycles, cars, and—increasingly—housing and home appliances and cooking, errands, and other household tasks.[55] Although some of the most visible of these new networks are for-profit enterprises (airbnb.com, Zipcar, taskrabbit.com; etc.), many others are organized as not-for-profit peer-to-peer initiatives.[56] These can be double-edged swords. Certainly, when they grow beyond the margins of economic relations some of these networks (especially the for-profit ones) pose serious dangers related to their circumvention of regulation and thereby the erosion of standards for safety, health, and income.[57] Yet to the extent that infrastructure or networks facilitate sharing of household responsibilities or allow more people to use fewer large appliances, autos, and other large items they also have great potential to lessen time poverty and resource consumption, as well as the sense that such roles are distinctly individual and private ones.

Home, then, is connected to citizenship of *some* sort. Recognizing this, we can attend more thoughtfully to the ways in which both household practices and the physical structuring of houses and the built environment itself can constrain or enable different manifestations of citizenship—both in general and for specific groups within these environments.

To argue that citizenship has a place in the home is to maintain that it is already present. Recognizing this presence is not necessarily to argue that it has a constructive place; it can also weaken or constrain public participation and discourse. These effects are also vital to the shaping of citizens and citizenship.

In both of these senses, the link between home and citizenship challenges familiar, arguably dominant, conceptualizations of purely *private* homes. It does little, however, to reconstruct notions of the participation of citizens. The focus has been on various ways in which citizen action is necessarily embedded in everyday practices of home and so upon the ways in which these practices *condition or shape*—as well as constrain or enable—citizenship. There is a related sense, however, in which home can be understood as integral to citizenship, but in this case we need to expand what we understand by citizenship and how we understand the home.

This third sense of citizenship emerges when we observe that the home can be, and often has been, a physical space for the exercise of citizenship itself: both as object and as subject. Sherilyn MacGregor draws this conclusion from her interviews with numerous activist women:

Activism is typically associated with activities in the public domain. Yet the association of activism and publicity (and the concomitant depoliticization of the private sphere) is challenged when women choose to regard household issues as political issues and thereby make their homes a focus for their activist engagement. Their homes are both the base for their public activism (i.e., "command central") and a place where political action and conversation takes place.[58]

Here home is a venue for citizenship. Even more significantly, MacGregor notes that it is also a *subject*—making political issues of household practices and material conditions. Historians Vanessa Taylor and Frank Trentmann offer different illustrations of household practices becoming the basis for citizenship in Victorian England. Here the demand for a continuous and affordable flow of water into the home became a central political demand of increasingly well-organized citizens.[59] As their study concludes: "Instead of bifurcating private and public, new practices— like running a bath or turning on the tap—came to channel political energy between these worlds. Conflicts over the legitimacy and scope of 'domestic' use had, by the end of the nineteenth century, broadened into a politics of entitlement and provision in times of scarcity. The material politics of everyday life played a vital role in expanding the mode of politics from taxation to provision."[60] Citizenship, here, engages the materiality of the household and emerges from our experience with this.[61] The focus on provision reflects the transformation of a social and material question into a political claim, reflecting Hanna Pitkin's argument that social questions must not be excluded from public life but must enter in the right "spirit."[62]

The possibilities for politicizing the material practices of the home are more expansive than we often recognize, in part because these practices are less fixed and subject to far more change than we often recognize. As Witold Rybczynski has observed, for example, the explicit notion of "comfort" as a goal or even criteria of the home and household does not appear to have emerged in Europe until the eighteenth century.[63] Relatedly, the notion of the home as a private haven would have made little sense in a society in which homes were also frequently regarded as places of work—for artisans or merchants, for example—and in which even bourgeois houses often contained more than one family.[64] Such dramatic changes are not just those that have taken place over centuries.

Standards of comfort, cleanliness, and convenience have changed dramatically over just the past several decades. Elisabeth Shove's research makes clear that "meanings of comfort and cleanliness do not represent free-floating expressions of personal preference."[65] As practices, they emerge at the confluence of innovation and diffusion of technologies, with evolving cultural norms regarding usage. For example, what is regarded as a "normal" and comfortable indoor air temperature has changed radically with the rise of central heating and air-conditioning and the concomitant changes toward year-round lightweight clothing in many places. Moreover, although diverse cultural criteria for comfort used to be commonplace, the trajectory is toward increasing uniformity.[66] Similar accounts have been offered of dramatic change in other household practices, including bathing or showering and clothes washing. Perhaps most striking in these accounts is the ways in which change has occurred not simply in terms of frequency or quantity (e.g., bathing times per week or pounds of laundry washed) but in terms of the very purposes of these activities (for example, whether clothes need to be washed in order to remove dirt or stains from the outside, to remove sweat or bodily odors from the inside, to disinfect, or to refresh the shape or fit).[67] Simply promoting greater resource efficiency of technology will not alter these practices and can simply encourage more usage. This leads Shove to conclude that cultivating a greater diversity of understandings of comfort, convenience, and cleanliness is among the most significant changes that can reduce the resource consumption of household practices.[68] It is in this sense that "green" practices might open these up to differing perspectives on their importance, a sense that they are not "merely" private practices and choices but ones that have important public consequences. There is increasing empirical evidence that this is the case—that rather than thinking of practices of consumption as either equivalent to or an alternative to practices of citizenship, they are instead complementary.[69]

Finally, it is a characteristic of modern homes that although conceived as a private sphere identified with materiality this very materiality is nonetheless hidden from view there. Maria Kaika has shown convincingly that the modern home is conceived "through a dual practice of exclusion: through ostracizing the undesired social as well as the undesired natural elements and processes."[70] Socially, she notes, homes are envisioned to exclude all undesired or uninvited others. Naturally, they are envisioned as excluding "dust, cold or polluted air, rain, dirt, sewerage, smog, etc."[71] These material forms of exclusion—through door locks and adequate roofing, for example—are key to our very idea of home. Yet the engineering of the complex urban infrastructure that distinguishes modern homes is designed to provide a flow of water, gas, and electricity into the home and systems of sewage and sanitation for the material flow of "waste" out of the home, as well as new standards of public health and expectations of cleanliness.[72]

Although forms of exclusion may be central to all homes, the deep dependence of the material construction of the home upon social and natural processes is particularly hidden in the modern home. As Kaika notes, the nature that is excluded from the home is a "bad" nature—dust, dirt, insects, rain, and cold. Yet the home is reliant upon the inclusion of a "good" nature—clean water, fresh air, various forms of energy. Kaika observes, "the function of the modern home as safe and autonomous is predicated not only upon the exclusion of bad nature from its premises, but also upon the visual exclusion of the networks and social relations that produce and transport good nature into the domestic and pump bad nature back into the urban domain."[73] We conceive the home as a boundary between inside and outside, but the often unacknowledged reality is that it is "a porous membrane" that controls the interaction between these two.[74] Indeed, for many, feeling at home relies upon "remaining unfamiliar with the socio-natural networks that produce domesticity."[75]

In this context, explicitly attending to these socionatural networks—whether by choice or necessity—can be the basis for a citizenly engagement and participation with the world in which the home is situated. Kaika concludes: "Demonstrating the ideological construction of private spaces as autonomous and disconnected and insisting on their material and social connections calls for an end to individualization, fragmentation and disconnectedness that are looked for within the bliss of one's home. It calls for engaging in political and social action, which is, almost invariably, decidedly public."[76] The awareness cultivated by attention to the materiality of the home—how and how much energy and water is consumed and for what purposes, what products are brought into the home and how much of it is regenerated as waste, how this waste is disposed of—can lead not only to changes in household practices but to the sort of intimate and experiential understanding of these material flows that can inform and prompt broader forms of collective action.

None of this makes sense unless it is clear that the everyday material practices of the household—care, provisioning, social reproduction, and so on—cannot be reduced to the role of a consumer.[77] These can be the basis for collaborative action rather than merely the expression of individual consumer preferences. Although the former creates opportunities for easing and improving the quality of household practices and so everyday life, the latter entails choices that aim primarily to assuage a sense of individual guilt or responsibility. The mere aggregation of "simple" individual acts by self-conscious consumers will not affect the large-scale social changes necessary, and relabeling this as citizenship does nothing to change this. However, if instead we see the ongoing engagement with "greening" the home as challenging the sense of inevitability and naturalism of household practices, then these practices can play a significant role in foregrounding and thus politicizing and restructuring these everyday material relations. Moreover, by recognizing home dwelling as a practice, we can understand it in ways that transcend the individualism of a purely market-based, consumer-oriented response. This opens up new opportunities for these practices to become the subject of collaborative or collective citizen attention and action.

In sum, the materiality of home, in all the manifestations discussed here, shapes, constrains, and enables the participation and citizenship of household members. Although authors including Winner and Maniates are quite right to warn against regarding individual consumer choices as a sufficient basis for constructive social change, we are not thereby justified in ignoring the ways in which the supposedly private household both shapes and constrains the opportunity for public citizen action. Citizenship is at stake in the home, whether "green" practices are employed or not. By recognizing this, we can imagine and create opportunities for citizens to politicize home practices and thereby explore and experiment with new ways of living well.

8

Conclusion

My aim in this book has been to pry open space for a more broadly and deeply resonant form of environmental criticism, which is thereby more democratic and ultimately more hopeful. The path to such resonance, I have argued, takes us into the thicket of everyday practices, and attention to these practices requires us to recognize their inescapable materiality. If there is merit in this project, it is not in that I have called for or outlined something wholly new—I have not—but in that I have drawn attention to the significance of some existing ideas, practices, and movements while casting them in a new light. This claim is no false modesty on my part: a central premise of this book has been the pragmatic conviction that theoretical insight can best be generated by attending to material practice and lived experience and that theories and ideologies formulated apart from these are more likely to lead us astray than to illuminate contemporary challenges.

Beyond Pessimism and Optimism

To assert that a more resonant form of environmental criticism is both more democratic and hopeful is to contrast this with familiar tropes in environmental argument that are neither. The most obvious of these is the pessimistic sort of catastrophism that imagines experts—especially scientists—as providing a top-down basis for action to "save us" from the climate crisis or other ecological disasters. This sort of catastrophism is often characterized as a legacy from 1970s-era thinking about environmental crisis.[1] Yet although this line of argument has been roundly criticized by critical intellectuals within the academy it never really vanished from broader popular discourse. Indeed, there is good reason to see it as resurgent in the face of contemporary concern about climate crisis. It is an argument that doesn't simply characterize environmental challenges in apocalyptic terms but explicitly links these challenges to a particular prescription for social change that is presented as having been derived from scientific or technical analysis.[2] For some, crisis itself is anticipated as an autonomous trigger for necessary action taken in the absence of choice or freedom. For others, the *threat* of catastrophe might become so real and imminent that people will "see the light" and then be convinced to do what is deemed by experts to be necessary. The "hard" version of this argument is as uncompromising as ever— technocratic authoritarianism as the only feasible alternative to the failure of democracy.[3] Yet it is a "softer" version that often seems more alluring and so is more widespread: a sense that the only way to provoke needed change is to shout louder, to shake people of their ignorance and complacency, or wait for crisis to hit them directly. The weakness of such an approach is not—as climate denialists and their fellow travelers would have it—to be found in the science per se but in the role that this approach attributes to science in fostering change. This sort of outside criticism doesn't so much reject the resonance dilemma as it abandons or writes off the possibility for more resonant criticism, deferring instead to experts or crisis to force the issue.

The opposite of pessimism is optimism. In this case, the opposite of the pessimistic catastrophism I have sketched is the approach often discussed under the label of "ecological modernization" in Europe and captured in US works, including Paul Hawken, Amory Lovins, and Hunter Lovins' *Natural Capitalism* and William McDonough and Michael Braungart's *Cradle to Cradle*.[4] According to this view, the same technological changes needed to avert crisis will also, happily, make all our lives more comfortable and enjoyable. The opening pages of *Natural Capitalism* make this point vividly:

Imagine . . . a world where cities have become peaceful and serene because cars and buses are whisper quiet, vehicles exhaust only water vapor, and parks and greenways have replaced unneeded urban freeways . . . Industrialized countries have reduced resource use by 80 percent while improving the quality of life . . . Is this the vision of a utopia? *In fact, the changes described here could come about in the decades to come as the result of economic and technological trends already in place.*[5]

The key point in this quotation is that desirable change is characterized as entirely dependent upon "economic and technological trends already in place." It is because they are convinced of the extant power of these currents that they assert that their imagined world is neither naïve nor utopian but likely. This view of economy and technology as autonomous displaces both the need for and the value of a more vigorous conception of democratic citizenship or social movements for change.[6] The task at hand is avowedly instrumental or procedural; "It is neither conservative nor liberal in its ideology . . . since it is a means, and not an end."[7]

This sort of approach has been infused with new rhetorical intensity in writing by activists Michael Shellenberger and Ted Nordhaus, Stewart Brand, and academics including Rasmus Karlsson.[8] As Karlsson puts it, we must pursue "the development of advanced technologies that would allow humanity to transcend its planetary boundaries."[9] There is clearly an argument for the need to change here, but the change envisioned is largely restricted to the technological and energy sources that serve as the instrumental means to extant social ends. Explicitly or implicitly, the argument is directed narrowly toward economic and political elites who have the power to redirect resources and investments to facilitate these transformative technical innovations. In this sense, the approach is that of an inside player, persuading those with this power to effect this redirection. The broader public message seems to be—to borrow a phrase— "keep calm and carry on."

This optimistic modernism seems most contradictory on the question of what obstacles are impeding such innovation now. The most explicit answer can be seen in these authors' often vehement criticisms of existing environmental organizations and arguments.[10] Yet given the elite-driven and technocratic nature of their proposed solutions, it is not at all clear why these organizations and arguments should be such a colossal impediment.

In one sense, clearly, pessimistic catastrophism and optimistic modernism appear as opposites. Yet they share significant characteristics that distinguish both from the approach I have pursued in this book. First, they both give overwhelming normative weight to the influence of experts and expertise at the expense of the broader populace. As a result, neither pursue a critical engagement with everyday experience and the material practices integral to it. Second, both treat the resonance dilemma as an unchangeable fact. For the pessimists, the lack of resonance will only be overcome when crisis trumps the everyday; for optimists the lack of resonance need not be overcome, because existing lifestyles and concepts of development can simply be fueled by new energy sources and technologies.

In the end, it is *hope* that stands in contrast to both pessimism *and* optimism. It is grounded in neither the optimist's apparent confidence that things will turn out for the best nor the pessimist's cynicism in giving up on the possibilities for change from within. As I've argued at greater length elsewhere, hope is the basis for action in a world that is evidently not the best of all possible ones.[11] Environmental justice activists, for example, are far too intimately familiar with racism, marginalization, and injustice to be optimists. Things do not simply "work out for the best." Yet despite the evidence that history and daily experience might seem to offer for pessimism and resignation, their activism challenges that. The action itself—reflective of existing and often deeply rooted concern for the people and places closest to one—is a democratic expression of hope. Democracy here is not reflective of a particular institutional arrangement or process but a quality reflected in the action itself. It is a manifestation of citizenship and inside criticism. As such, it helps us imagine and open up possibilities for change from within the material practices of everyday life.

My chapters on land, automobility, and homes are intended to suggest the sorts of resources and perspectives that might help us to cultivate a more critical and expansive imaginary. To do so requires taking material practices seriously and recognizing the diverse ways in which key political values are embedded with these practices. I have no doubt that others—those with experiences that I have not anticipated here and scholars with deeper knowledge of these practices—may challenge my reading of the particular way that (for example) freedom does and might relate to automobility or that citizenship can and should emerge from within the practices in the home. I would take this, however, as a reflection of success for the project as a whole. The conversation that could ensue would be on the broad but underdeveloped terrain of inside criticism—a hopeful development indeed.

Shifting the Terrain of Environmental Social Criticism

My aim has been to make the case for a distinct way of *doing* environmental criticism. Rather than beginning with liberalism, modernity, or other grand theoretical traditions, we must begin with respect for the complexity and sincerity of people's values and everyday experience. Criticism must be closely engaged with these. As such, there can be no such single thing as a "resonant" environmentalism, because the plurality of practices and contexts within which our lives are embedded ensures that contest and controversy are inescapable. Yet criticism that seeks resonance demands the critic's attention to our material practices and contexts; it doesn't promise final settlement but does offer the prospect of greater agreement upon means and identification of particular, shared ends.

Both obstacles and opportunities emerge from this attention. On the one hand, we can see in a more fine-grained manner the ways in which engagement with existing material practices enables a sense of convenience, privacy, identity, or freedom that is meaningful to many. I have sought to provide illustrations of this, on different scales, in previous chapters. A reused plastic grocery bag eases a parent's hectic morning by providing a quick way to transport a wet swimsuit; land ownership provides some with privacy and security; a car facilitates flexible mobility and expands options for employment, social engagement, or escape from parental surveillance; household appliances are "labor-saving" devices for many. Stated plainly, such observations appear banal. Yet we must recognize that arguments against such conveniences can be deeply threatening, because, as Rousseau long ago recognized, "being deprived of them [is] much more cruel than possessing them was sweet."[12] Moreover, beginning with real recognition of such experiences is a precondition for criticism that minimizes the chances of being patronizing or arrogant.

On the other hand, critical attention to everyday practices simultaneously allows us to recognize the particular ways in which the inability to drive a car in a community structured by automobility constrains the young, the old, and the carless, whereas the near necessity of driving and owning a car is an imposition upon many others. It allows us to see the ways in which the structure of private households and neighborhoods places a disproportionate burden on the social reproductive work of women. It illuminates always already present constituencies, spaces, and ways in which criticism can resonate by engaging openly with differences in identity and circumstance, including class, gender, race, and other markers of relational social position. In this sense, it can direct our attention toward the many ways in which practice is at odds with, or falls short of, influential theories and conceptualizations. These gaps are where resonant criticism can thrive.

In the end, the critical and reconstructive aspects of the argument here aim for the same target: a more expansive conversation that resonates more broadly and deeply because it is attuned to the concerns, frustrations, pleasures, and fears that animate us every day. It is not my aim to propose a list of policy or institutional changes that ought to follow from this. These will vary based upon context and place. Yet I do gesture toward some approaches that seem promising in this regard.

The proclivity to engage the everyday can be found in a variety of recent movements, theoretical approaches, and academic initiatives. Environmental and climate justice movements, theories of a "new materialism," and projects to reframe environmentalism and "climate change communication" are a few of the ones I have addressed throughout this book. By focusing on the qualities necessary for a more resonant environmentalism, I have sought to identify ways in which these qualities are found in such places and ways in which they are not.

For instance, to the extent that a "new materialism" in social and political theory pushes us to engage with and reflect upon these material practices of the everyday, it is a promising resource. When matter is recognized as vibrant, not "dead"; interconnected, not self-contained; enmeshed with human values, not separated from them, then material practices can be given their due. Yet I have argued that the preoccupation with ontology in many such works supposes we must (and can) transform our way of seeing the world in order to engage its materiality. We don't need to, and calls to do so work at cross-purposes with the promise of engaging the everyday.

The materialism that we need is not really all that "new." Yet in affluent postindustrial societies it may seem new, in the sense that it stands in contrast to dominant ways of conceiving both environmentalism and politics. It demands attention to how we live and work and play. It invites us to think in more grounded ways about property and freedom and citizenship. Precisely because it does this, it invites us to imagine a society in which "better" need not mean "more" and in which environmental *politics* is the name we give to the plural and multifaceted struggle to build flourishing communities that might be sustained over time. This vision may seem inadequate in the face of multiple crises of climate, bio-diversity loss, and environmental toxins. It promises no silver bullet, no singular solution. It does not resolve questions such as how to balance demands for carbon mitigation with adaptation to a changing climate nor does it identify the best transportation or housing options for par-ticular communities or states. Yet if, as I have argued, the real crisis is the resonance dilemma, then I believe it can help us to see a path forward, which allows us to confront this more openly and honestly. In doing so, we move onto a terrain that allows us to grapple with these other crises in a more creative, effective, and hopeful manner.

Notes

1 Introduction

1. Michael D. Shear, "Obama Tells Donors of Tough Politics of Environment," *New York Times*, April 4, 2013, sec. U.S./Politics, http://www.nytimes.com/2013/04/05/us/politics/obama-donors-keystone-pipeline.html.

2. "Thirteen Years of the Public's Top Priorities," January 27, 2014, Pew Research Center for the People & the Press, http://www.people-press.org/interactives/top-priorities/; Matthew C. Nisbet, "Public Opinion and Participation," in *Oxford Handbook of Climate Change and Society*, ed. John S. Dryzek, Richard B. Norgaard, and David Schlosberg (Oxford: Oxford University Press, 2011), 355–368; Deborah L. Guber, *The Grassroots of a Green Revolution: Polling America on the Environment* (Cambridge, MA: MIT Press, 2003).

3. Christopher F. Schuetze, "Environmental Warning Fatigue Sets In," *IHT Rendezvous* (blog), *New York Times*, March 2, 2013, http://rendezvous.blogs.nytimes.com/2013/03/02/environmental-warning-fatigue-sets-in/.

4. Dorceta E. Taylor, "The State of Diversity in Environmental Organizations: Mainstream NGOs, Foundations & Government Agencies." *Green 2.0 Working Group*, July 2014, http://diversegreen.org/report/; Talia Buford, "Greens Confront Own Need for Diversity," *POLITICO*, December 29, 2012, http://www.politico.com/story/2012/12/greens-confront-own-need-for-diversity-85558.html; Brentin Mock, "Mainstream Green Is Still Too White," *COLORLINES*, April 2, 2013, http://colorlines.com/archives/2013/04/message_from_the_grassroots_dont_blow_it_on_climate_change_this_time.html.

5. Katie Valentine, "The Whitewashing of the Environmental Movement," *Grist*, September 24, 2013, http://grist.org/climate-energy/the-whitewashing-of-the-environmental-movement/; I develop this point in "Populism, Paternalism, and the State of Environmentalism in the U.S.," *Environmental Politics* 17, no. 2 (2008): 219–236.

6. Mock, "Mainstream Green Is Still Too White."

7. Christopher Ingraham, "The Green Movement Has a Millennial Problem," *Washington Post*, March 7, 2014, http://www.washingtonpost.com/blogs/wonkblog/wp/2014/03/07/the-green-movement-has-a-millennial-problem/.

8. E.g., Graham Smith, *Deliberative Democracy and the Environment* (London: Routledge, 2004); Thomas Prugh, Robert Costanza, and Herman E. Daly, *The Local Politics of Global Sustainability* (Washington, DC: Island Press, 2000). John Dryzek's body of work on discursive democracy is definitive here. See, for example, John S. Dryzek, "Green Democracy," in *Deliberative Democracy and Beyond: Liberals, Critics, Contestations* (Oxford: Oxford University Press, 2000); John S. Dryzek and Hayley Stevenson, "Global Democracy and Earth System Governance," *Ecological Economics* 70, no. 11 (September 2011): 1865–1874, doi:10.1016/j.ecolecon.2011.01.021.

9. Kersty Hobson, "On the Making of the Environmental Citizen," *Environmental Politics* 22, no. 1 (February 2013): 67, doi:10.1080/09644016.2013.755388.

10. Ibid.

11. Cf. Nicholas Lemann, "When the Earth Moved," *The New Yorker*, April 15, 2013, http://www.newyorker.com/arts/critics/atlarge/2013/04/15/130415crat _atlarge_lemann; Ingolfur Blühdorn has argued that this political context is no longer feasible and therefore that the democratic action I envision as a possibility here is doomed to fail. Although his analysis illuminates the contemporary "politics of unsustainability," my project here is to explore avenues that Blühdorn's more deterministic analysis might foreclose: Ingolfur Blühdorn, "The Governance of Unsustainability: Ecology and Democracy after the Post-Democratic Turn," *Environmental Politics* 22, no. 1 (February 2013): 17, doi:10.1080/0964 4016.2013.755005; see also Ingolfur Blühdorn, "The Politics of Unsustainability: COP15, Post-Ecologism, and the Ecological Paradox," *Organization & Environment* 24, no. 1 (March 1, 2011): 34–53, doi:10.1177/1086026611402008.

12. Given my use of the language of "resonance," it is worth noting that William Connolly's account of an "evangelical-capitalist resonance machine" has more in common with this alternate diagnosis, focused on the virulence of opposition. William E. Connolly, "The Evangelical-Capitalist Resonance Machine," *Political Theory* 33, no. 6 (December 2005): 869–886, doi:10.1177/0090591705280376.

13. Riley E. Dunlap and Aaron M. McCright, "Organized Climate Change Denial," in *Oxford Handbook of Climate Change and Society*, ed. John S. Dryzek, Richard B. Norgaard, and David Schlosberg (Oxford: Oxford University Press, 2011): 144–160.

14. Canada and Australia seem to be the exceptions here.

15. Nisbet, "Public Opinion and Participation," 359.

16. "Two-thirds of Americans (67%) say there is solid evidence that the earth has been getting warmer over the last few decades, a figure that has changed little in the past few years." Pew Research Center for the People and the Press, "GOP Deeply Divided over Climate Change," November 1, 2013, http://www .people-press.org/files/legacy-pdf/11-1-13%20Global%20Warming%20Release .pdf.

17. "Thirteen Years of the Public's Top Priorities," 8.

18. Mike Hulme, *Why We Disagree about Climate Change: Understanding Controversy, Inaction, and Opportunity* (Cambridge: Cambridge University Press, 2009).

19. For further development of this, see my essay "Populism, Paternalism, and the State of Environmentalism in the U.S."

20. William Chaloupka, "The Tragedy of the Ethical Commons: Demoralizing Environmentalism," in *The Politics of Moralizing*, ed. Jane Bennett and Michael J. Shapiro (New York: Routledge, 2002), 121.

21. Richard Flacks, *Making History: The American Left and the American Mind* (New York: Columbia University Press, 1988), 237; yet as Michael Walzer argues persuasively, the biblical prophets rarely fit this "outsider" model: Michael Walzer, *Interpretation and Social Criticism* (Cambridge, MA: Harvard University Press, 1993).

22. Derrick Jensen, *Endgame, Volume 1: The Problem of Civilization* (New York: Seven Stories Press, 2005), ix–xii.

23. Flacks, *Making History*, 237.

24. Jane Bennett, "The Moraline Drift," in *The Politics of Moralizing*, ed. Jane Bennett and Michael J. Shapiro (New York: Routledge, 2002), 11–26.

25. An analysis of the attractions and dangers of this sort of thinking can be found in Sasha Lilley et al., *Catastrophism: The Apocalyptic Politics of Collapse and Rebirth* (Oakland, CA: PM Press, 2012).

26. Walzer, *Interpretation and Social Criticism*, 31.

27. James Tully, *Public Philosophy in a New Key*, vol. I, *Democracy and Civic Freedom* (Cambridge: Cambridge University Press, 2008), 15–38; David Forgacs, ed., *An Antonio Gramsci Reader* (New York: Schocken Books, 1988), 300–311.

28. Flacks, *Making History*, 92.

29. On total revolution, see Bernard Yack, *The Longing for Total Revolution: Philosophic Sources of Social Discontent from Rousseau to Marx and Nietzsche* (Princeton, NJ: Princeton University Press, 1986); on worldview transformation among radical environmental thinkers, see my *Political Nature: Environmentalism and the Interpretation of Western Thought* (Cambridge, MA: MIT Press, 2001), 21–34.

30. The slogan of the activist group Earth First!, for instance, is "No Compromise in Defense of Mother Earth."

31. Marcel Wissenburg has recently formulated a three-fold categorization of critique as "practical," "fundamental," or "radical," which shares some characteristics, respectively, with my "inside players," "inside critics," and "outside critics." See Marcel Wissenburg, "Political Appeasement and Academic Critique: The Case of Environmentalism," *Philosophy & Social Criticism* 39, no. 7 (2013): 684, doi:10.1177/0191453713491233.

32. Val Plumwood, "Inequality, Ecojustice, and Ecological Rationality," in *Debating the Earth: The Environmental Politics Reader*, ed. John S. Dryzek and David Schlosberg (Oxford: Oxford University Press, 1998), 559–583; Mark B. Brown, *Science in Democracy: Expertise, Institutions, and Representation* (Cambridge, MA: MIT Press, 2009); Kerry H. Whiteside, "The Impasses of Ecological

Representation," *Environmental Values* 22, no. 3 (June 1, 2013): 339–358, doi: 10.3197/096327113X13648087563700.

33. For an overview, see my "Political Theory and the Environment," in *Oxford Handbook of Political Theory*, ed. John S. Dryzek, Bonnie Honig, and Anne Phillips (Oxford: Oxford University Press, 2006); also Teena Gabrielson et al., eds., *Oxford Handbook of Environmental Political Theory* (Oxford: Oxford University Press, forthcoming 2015).

34. Michael Walzer, "The Political Theory License," *Annual Review of Political Science* 16, no. 1 (2013): 1–9, doi:10.1146/annurev-polisci-032211-214411. His characterization of this license also emphasizes the freedom of political theorists to take substantive political positions in their work.

35. I develop this point further in "Political Theory and the Environment," 775–777.

36. Two insightful critiques of this sort of treatment of modernity are Bernard Yack, *The Fetishism of Modernities: Epochal Self-Consciousness in Contemporary Social and Political Thought* (Notre Dame, IN: University of Notre Dame Press, 1997); and Bruno Latour, *We Have Never Been Modern* (Cambridge, MA: Harvard University Press, 1993).

37. John Dewey, *Experience and Nature* (George Allen & Unwin Ltd., 1929), 7.

38. Among social theorists, Anthony Giddens offers perhaps the clearest account of this conceptualization of practice: *The Constitution of Society: Introduction of the Theory of Structuration* (Oakland: University of California Press, 1984), 1–40; more recently, a number of social scientists and philosophers have sought to advance "practice theory" as an empirical approach to understanding social change and stability. The following provide relevant and helpful overviews: Elizabeth Shove, Mika Pantzar, and Matt Watson, *The Dynamics of Social Practice: Everyday Life and How It Changes* (London: Sage, 2012), 1–19; Gert Spaargaren, "Theories of Practices: Agency, Technology, and Culture," *Global Environmental Change* 21, no. 3 (August 2011): 813–822, doi:10.1016/j.gloenvcha.2011.03.010; Alan Warde, "Consumption and Theories of Practice," *Journal of Consumer Culture* 5, no. 2 (July 1, 2005): 131–153, doi:10.1177/1469540505053090.

39. On this theme, see Michael Maniates and John M. Meyer, eds., *The Environmental Politics of Sacrifice* (Cambridge, MA: MIT Press, 2010).

40. Manuel Arias-Maldonado, *Real Green: Sustainability after the End of Nature* (Farnham: Ashgate, 2012), 118.

41. Rasmus Karlsson, "Individual Guilt or Collective Progressive Action? Challenging the Strategic Potential of Environmental Citizenship Theory," *Environmental Values* 21, no. 4 (November 1, 2012): 464, doi:10.3197/0963271 12X13466893628102; cf. Ted Nordhaus and Michael Shellenberger, *Break Through: From the Death of Environmentalism to the Politics of Possibility* (Boston: Houghton Mifflin, 2007).

42. John M. Meyer, "A Democratic Politics of Sacrifice?," in *The Environmental Politics of Sacrifice*, ed. Michael Maniates and John M. Meyer (Cambridge, MA: MIT Press, 2010), 21.

43. Christopher Buck identifies our contemporary anxiety about a time-crunch as such a "space" in "Post-Environmentalism: An Internal Critique," *Environmental Politics* 22, no. 6 (November 2013): 883–900, doi:10.1080/09644016.2 012.712793.

44. It is important to recognize that a considerable portion of climate change emissions and other ecological impacts in China, India, and elsewhere are also a product of facilities in these countries that produce for consumption in wealthy, postindustrial societies.

45. Andrew Dobson, *Citizenship and the Environment* (Oxford: Oxford University Press, 2003), 119.

46. Ramachandra Guha and Juan Martinez-Alier, *Varieties of Environmentalism: Essays North and South* (London: Earthscan, 1997).

47. For an exception, see Julian Agyeman, *Introducing Just Sustainabilities: Policy, Planning and Practice* (London: Zed Books, 2013).

2 We Have Never Been Liberal

1. I will complicate this understanding of liberalism later in the chapter. My point here, however, is to distinguish liberalism understood as a broad philosophy from a narrower view of liberalism as a political tradition—whether the US-based connotation of a progressive, social-welfare position or in the continental European sense of a more market-based libertarianism.

2. Ruth Abbey, "Is Liberalism Now an Essentially Contested Concept?," *New Political Science* 27, no. 4 (December 2005): 462, doi:10.1080/07393140500370972; see also John S. Dryzek, Bonnie Honig, and Anne Phillips, "Editors' Introduction," in *Oxford Handbook of Political Theory* (Oxford: Oxford University Press, 2006), 23–24.

3. Marcel Wissenburg, *Green Liberalism* (London: UCL Press, 1998), 3.

4. Simon Hailwood, *How to Be a Green Liberal: Nature, Value and Liberal Philosophy* (Chesham, UK: Acumen, 2004); I discuss this further in John M. Meyer, "Green Liberalism and Beyond," *Organization and Environment* 18, no. 1 (March 2005): 116–120; cf. Derek Bell, "How Can Political Liberals Be Environmentalists?," *Political Studies* 50, no. 4 (2002): 703–724, doi:10.1111/1467-9248.00003.

5. Although an overlapping consensus (in Rawls's sense) might justify a commitment to freedom, tolerance, and participation, for example, it is hard to see how any such consensus could persuade others to address environmental concerns with a commitment to something as amorphous to the general public as "liberalism" per se.

6. Richard Dagger, "Freedom and Rights," in *Political Theory and the Ecological Challenge*, ed. Andrew Dobson and Robyn Eckersley (Cambridge: Cambridge University Press, 2006), 200–201.

7. Abbey, "Is Liberalism Now an Essentially Contested Concept?," 461.

8. John Gray, "Back to Mill: Review of *The Snake That Swallowed Its Tail: Some Contradictions in Modern Liberalism* by Mark Garnett," *The New Statesman* 17, no. 833 (2004): 50.

9. Louis Hartz, *The Liberal Tradition in America: An Interpretation of American Political Thought since the Revolution*, (New York: Harcourt Brace, 1955), 177.

10. Marcel Wissenburg, "Liberalism Is Always Greener on the Other Side of Mill: A Reply to Piers Stephens," *Environmental Politics* 10, no. 3 (2001): 24.

11. Raymond Geuss, "Liberalism and Its Discontents," *Political Theory* 30, no. 3 (June 2002): 320.

12. On the difference between the practice of toleration and an attitude of tolerance, see Andrew R. Murphy, "Tolerance, Toleration, and the Liberal Tradition," *Polity* 29, no. 4 (1997): 593.

13. Eduard Bernstein, *Evolutionary Socialism* (New York: B. W. Huebsch, 1911), 206.

14. Cf. Marx's preface to the first German edition of Capital in Karl Marx and Friedrich Engels, *The Marx-Engels Reader*, ed. Robert C. Tucker (New York: W.W. Norton, 1978), 298.

15. Cf. Abbey, "Is Liberalism Now an Essentially Contested Concept?"

16. Peter Hay, *A Companion to Environmental Thought* (Edinburgh: Edinburgh University Press, 2002), 195.

17. William Ophuls, *Ecology and the Politics of Scarcity: Prologue to a Political Theory of the Steady State* (San Francisco: W. H. Freeman, 1977).

18. Ophuls's three major books can be read as something of a trilogy. In addition to his 1977 work and the 1997 book under discussion here, see *Plato's Revenge: Politics in the Age of Ecology* (Cambridge, MA: MIT Press, 2011).

19. William Ophuls, *Requiem for Modern Politics: The Tragedy of the Enlightenment and the Challenge of the New Millennium* (Boulder, CO: Westview Press, 1997), ix.

20. Ibid., 1.

21. For further development of the significance of this total critique, see John M. Meyer, "Review of *Requiem for Modern Politics: The Tragedy of the Enlightenment and the Challenge of the New Millennium*, by William Ophuls," *Political Science* 51, no. 1 (July 1999): 51.

22. Joel J. Kassiola, *The Death of Industrial Civilization: The Limits to Economic Growth and the Repoliticization of Advanced Industrial Society* (Albany: SUNY Press, 1990), 83.

23. Ibid.

24. Ibid., 89.

25. Ibid., 86–87.

26. Andrew Vincent, "Liberalism and the Environment," *Environmental Values* 7 (1998): 443–444.

27. Andrew Dobson, *Green Political Thought*, second ed. (London: Routledge, 1995), 132.

28. Cf. Bernard Yack, *The Longing for Total Revolution: Philosophic Sources of Social Discontent from Rousseau to Marx and Nietzsche* (Princeton, NJ: Princeton University Press, 1986).

29. Terry L. Anderson and Donald R. Leal, *Free Market Environmentalism*, Revised ed. (New York: Palgrave, 2001), 26.

30. Cf. Mark Pennington, "Classical Liberalism and Ecological Rationality: The Case for Polycentric Environmental Law," *Environmental Politics* 17, no. 3 (2008): 431–448; Tom Bethell, *The Noblest Triumph: Property and Prosperity through the Ages* (New York: St. Martin's Griffin, 1998); Tibor R. Machan, "Pollution and Political Theory," in *Earthbound: New Introductory Essays in Environmental Ethics*, ed. Tom Regan (New York: Random House, 1984), 74–106.

31. Gregory S. Alexander, "Propriety through Commodity? Why Have Legal Environmentalists Embraced Market-Based Solutions?," in *Private Property in the 21st Century: The Future of An American Ideal*, ed. Harvey M. Jacobs (Cheltenham, UK: Edward Elgar, 2004), 75–91; Carol M. Rose, "The Several Futures of Property: Of Cyberspace and Folk Tales, Emission Trades and Ecosystems," *Minnesota Law Review*, no. 83 (1998): 129–182.

32. See Brian Barry, "Sustainability and Intergenerational Justice," in *Fairness and Futurity: Essays on Environmental Sustainability and Social Justice*, ed. Andrew Dobson (Oxford: Oxford University Press, 1999), 93–117; John Barry, "Greening Liberal Democracy: Practice, Theory and Political Economy," in *Sustaining Liberal Democracy*, ed. John Barry and Marcel Wissenburg (New York: Palgrave, 2001), 59–80; Bell, "How Can Political Liberals Be Environmentalists?"; Gideon Calder and Catriona McKinnon, "Introduction: Climate Change and Liberal Priorities," *Critical Review of International Social and Political Philosophy* 14, no. 2 (2011): 91–97, doi:10.1080/13698230.2011.529702; Avner de-Shalit, *The Environment between Theory and Practice* (Oxford: Oxford University Press, 2000); Gus diZerega, "Empathy, Society, Nature and the Relational Self: Deep Ecology and Liberal Modernity," *Social Theory and Practice* 21 (Summer 1995): 239–269; Andrew Dobson, *Citizenship and the Environment* (Oxford: Oxford University Press, 2003); Hailwood, *How to Be a Green Liberal: Nature, Value and Liberal Philosophy*; Mika LaVaque-Manty, *Arguments and Fists: Political Agency and Justification in Liberal Theory* (New York: Routledge, 2002); Brian G. Norton, "Ecology and Opportunity: Intergenerational Equity and Sustainable Options," in *Fairness and Futurity: Essays on Environmental Sustainability and Social Justice*, ed. Andrew Dobson (Oxford: Oxford University Press, 1999), 118–150; G. J. Paton, *Seeking Sustainability: On the Prospect of an Ecological Liberalism* (London: Routledge, 2011); Piers H. G. Stephens, "Green Liberalisms: Nature, Agency and the Good," *Environmental Politics* 10,

no. 3 (September 2001): 1–22; Steve Vanderheiden, *Atmospheric Justice: A Political Theory of Climate Change* (Oxford: Oxford University Press, 2008); Wissenburg, "Liberalism Is Always Greener on the Other Side of Mill: A Reply to Piers Stephens"; Wissenburg, *Green Liberalism*.

33. Barry, "Greening Liberal Democracy: Practice, Theory and Political Economy," 66.

34. Ibid., 67.

35. Stephens, "Green Liberalisms: Nature, Agency and the Good," 43.

36. Wissenburg, *Green Liberalism*, 213; Wissenburg, "Liberalism Is Always Greener on the Other Side of Mill: A Reply to Piers Stephens," 194.

37. Abbey, "Is Liberalism Now an Essentially Contested Concept?," 476.

38. Ophuls, *Requiem for Modern Politics: The Tragedy of the Enlightenment and the Challenge of the New Millennium*, 27.

39. Stephens, "Green Liberalisms: Nature, Agency and the Good," 5; see also Michael Freeden, *Liberal Languages: Ideological Imaginations and Twentieth-Century Progressive Thought* (Princeton, NJ: Princeton University Press, 2005), 11.

40. de-Shalit, *The Environment between Theory and Practice*, 83.

41. Although writing from outside what he took to be the liberal tradition, Karl Polanyi's work is a classic source for the development of this insight. I also address the relevance of Polanyi's thought in chapter 4.

42. Dobson, *Citizenship and the Environment*, 161.

43. Ibid.

44. Ibid.

45. Wissenburg, *Green Liberalism*, 123.

46. Marcel Wissenburg, "Sustainability and the Limits of Liberalism," in *Sustaining Liberal Democracy*, ed. John Barry and Marcel Wissenburg (New York: Palgrave, 2001), 201.

47. Robyn Eckersley, *The Green State: Rethinking Democracy and Sovereignty* (Cambridge, MA: MIT Press, 2004), 104.

48. Norton, "Ecology and Opportunity: Intergenerational Equity and Sustainable Options," 118; See also Brian G. Norton, *Sustainability: A Philosophy of Adaptive Ecosystem Management* (Chicago: University of Chicago Press, 2005), especially chapters 8 and 9.

49. Norton, "Ecology and Opportunity: Intergenerational Equity and Sustainable Options," 122.

50. Ibid., 119.

51. Ibid.

52. Ibid., 142.

53. Ibid.

54. Dobson, *Citizenship and the Environment*, 166.

55. John Barry and Marcel Wissenburg, *Sustaining Liberal Democracy: Ecological Challenges and Opportunities* (New York: Palgrave, 2001), 1.

56. Dobson, *Citizenship and the Environment*, 3.

57. Ibid., 111–114.

58. Ibid., 89.

59. Ibid., 139.

60. Ibid., 140.

61. Ibid., 141–142.

62. Barry, "Greening Liberal Democracy: Practice, Theory and Political Economy," 60.

63. Eckersley, *The Green State: Rethinking Democracy and Sovereignty*, 107.

64. See Mark Sagoff, "Can Environmentalists Be Liberals?," in *The Economy of the Earth: Philosophy, Law and the Environment* (Cambridge: Cambridge University Press, 1988), 146–170. Sagoff borrows his usage of this term from Bruce A. Ackerman, *Private Property and the Constitution* (New Haven, CT: Yale University Press, 1977).

65. Mark Sagoff, *The Economy of the Earth: Philosophy, Law and the Environment* (Cambridge: Cambridge University Press, 1988), 166; liberal advocates of toleration and diversity also offer a constrained vision of liberalism's ability to structure citizens' lives and group identities. For a preliminary effort to link these concerns, see Breena Holland, "Ecological Constraints and Value-Pluralism: Why Democracies Should Prohibit Some Ways of Life" (paper presented at the Western Political Science Association annual meeting, Oakland, CA, 2005).

66. I reexamine Sagoff's notion of a public–private divide and the related citizen–consumer dichotomy in chapters 4 and 7.

67. Sagoff, *The Economy of the Earth*, 167.

68. Cf. Zev Trachtenberg, "Complex Green Citizenship and the Necessity of Judgment," *Environmental Politics* 19, no. 3 (2010): 339–355.

3 The Question of Materiality in Environmental Politics

1. Giovanna Di Chiro, "Living Environmentalisms: Coalition Politics, Social Reproduction, and Environmental Justice," *Environmental Politics* 17, no. 2 (2008): 276–298, doi:10.1080/09644010801936230; David Schlosberg, "Theorizing Environmental Justice: The Expanding Sphere of a Discourse," *Environmental Politics* 22, no. 1 (2013): 37–55; Ronald Sandler and Phaedra C. Pezzullo, eds., *Environmental Justice and Environmentalism: The Social Justice Challenge to the Environmental Movement* (Cambridge, MA: MIT Press, 2006).

2. Ramachandra Guha and Juan Martinez-Alier, *Varieties of Environmentalism: Essays North and South* (London: Earthscan, 1997).

3. Cheryl Hall, "What Will It Mean to Be Green? Envisioning Positive Possibilities without Dismissing Loss," *Ethics, Policy & Environment* 16, no. 2 (June

2013): 125–141, doi:10.1080/21550085.2013.801182; Matthew C. Nisbet, "Communicating Climate Change: Why Frames Matter for Public Engagement," *Environment*, April 2009, www.environmentmagazine.org/Archives/Back Issues/ March-April 2009/Nisbet-full.html.

4. Ronald Inglehart, *The Silent Revolution: Changing Values and Political Styles among Western Publics* (Princeton, NJ: Princeton University Press, 1977); Ronald Inglehart, *Culture Shift in Advanced Industrial Society* (Princeton: Princeton University Press, 1990); Ronald Inglehart, "Changing Values among Western Publics from 1970–2006," *West European Politics* 31, nos. 1–2 (2008): 130–146; on the relation to environmentalism, see especially Ronald Inglehart, "Public Support for Environmental Protection: Objective Problems and Subjective Values in 43 Societies," *PS: Political Science and Politics* 28, no. 1 (March 1995): 57–72.

5. Samuel P. Hays, *Conservation and the Gospel of Efficiency; the Progressive Conservation Movement, 1890–1920* (Cambridge, MA: Harvard University Press, 1959).

6. Elmo R. Richardson, *The Politics of Conservation: Crusades and Controversies, 1897–1913* (Berkeley: University of California Press, 1962); Karl Jacoby, *Crimes against Nature: Squatters, Poachers, Thieves, and the Hidden History of American Conservation* (Berkeley: University of California Press, 2001).

7. William Cronon, "The Trouble with Wilderness; Or, Getting Back to the Wrong Nature," in *Uncommon Ground: Toward Reinventing Nature* (New York: W. W. Norton, 1995), 73; Christopher C. Sellers has recently advanced a reinterpretation of the rise of US environmentalism that places far greater emphasis upon suburban residents' engagement with their immediate environments. This offers a challenge to the conventional narrative that seems consistent with the critique that I develop in this chapter. See Christopher C. Sellers, *Crabgrass Crucible: Suburban Nature and the Rise of Environmentalism in Twentieth-Century America* (Chapel Hill: University of North Carolina Press, 2012).

8. Claus Offe, "Challenging the Boundaries of Institutional Politics: Social Movements since the 1960s," in *Changing Boundaries of the Political*, ed. Charles S. Maier (New York: Cambridge University Press, 1987), 66–70; cf. Robyn Eckersley, "Green Politics and the New Class: Selfishness or Virtue?," *Political Studies* 37 (1989): 205–223.

9. Offe, "Challenging the Boundaries," 66. A suggestive illustration is provided by the world's first national "Green" political party, founded in New Zealand in 1972. They named themselves the "Values Party," seemingly confident that the contrast between values and the dominant political preoccupation with material economic interests meant that the *particular* values to be advanced could be left unspecified in their party's name. Christine Dann, "'From Earth's Last Islands': The Development of the First Two Green Parties New Zealand and Tasmania," n.d., http://www.globalgreens.org/earths-last-islands-development-first -two-green-parties-tasmania-and-new-zealand; "Blast from the Past - Values Party TV Ad - Extended Video - Video - 3 News," June 1, 2012, http://www.3news .co.nz/Blast-from-the-past---Values-Party-TV-ad/tabid/315/articleID/256413/ Default.aspx.

10. Abraham H. Maslow, "A Theory of Human Motivation," *Psychological Review* 50, no. 4 (July 1943): 370–396, doi:10.1037/h0054346.

11. Inglehart, "Public Support for Environmental Protection," 70.

12. Ibid., 62.

13. David I. Stern, "The Rise and Fall of the Environmental Kuznets Curve," *World Development* 32, no. 8 (2004): 1419–1439.

14. Ted Nordhaus and Michael Shellenberger, *Break Through: From the Death of Environmentalism to the Politics of Possibility* (Boston: Houghton Mifflin, 2007), 28.

15. Ibid., 52 and 39.

16. E.g., Samuel P. Hays, *Beauty, Health, and Permanence: Environmental Politics in the United States, 1955–1985* (Cambridge University Press, 1987) and Guha and Martinez-Alier, *Varieties of Environmentalism: Essays North and South*, xi–xv, trace the origins of the postmaterialist interpretation back much further.

17. E.g., First National People of Color Environmental Leadership Summit, "Principles of Environmental Justice," 1991, http://www.ejnet.org/ej/principles .html; Daniel Faber, *The Struggle for Ecological Democracy: Environmental Justice Movements in the United States* (New York: Guilford Press, 1998).

18. E.g., Bron R. Taylor, *Ecological Resistance Movements* (Albany: State University of New York Press, 1995); Guha and Martinez-Alier, *Varieties of Environmentalism: Essays North and South*.

19. Robert Gottlieb, *Forcing the Spring: The Transformation of the American Environmental Movement* (Washington DC: Island Press, 1993).

20. Riley E. Dunlap, G. H. Gallup, and A. M. Gallup, *Health of the Planet: Results of a 1992 International Environmental Opinion Survey of Citizens in 24 Nations* (Princeton, NJ: George H. Gallup International Institute, 1993).

21. Steven R. Brechin, "Objective Problems, Subjective Values, and Global Environmentalism: Evaluating the Postmaterialist Argument and Challenging a New Explanation," *Social Science Quarterly* 80, no. 4 (December 1999): 794.

22. Riley E. Dunlap and Richard York, "The Globalization of Environmental Concern and the Limits of the Postmaterialist Values Explanation: Evidence from Four Multinational Surveys," *Sociological Quarterly* 49, no. 3 (2008): 529–563; Jennifer E. Givens and Andrew K. Jorgenson, "The Effects of Affluence, Economic Development, and Environmental Degradation on Environmental Concern: A Multilevel Analysis," *Organization & Environment* 24, no. 1 (March 1, 2011): 74–91, doi:10.1177/1086026611406030; So Young Kim and Yael Wolinsky-Nahmias, "Cross-National Public Opinion on Climate Change: The Effects of Affluence and Vulnerability," *Global Environmental Politics* 14, no. 1 (February 2014): 79–106, doi:10.1162/GLEP_a_00215.

23. Andrew Dobson, *Justice and the Environment: Conceptions of Environmental Sustainability and Dimensions of Social Justice* (Oxford: Oxford University Press, 1998), 25; emphasis added.

24. Kim Allen, Vinci Daro, and Dorothy Holland, "Becoming an Environmental Justice Activist," in *Environmental Justice and Environmentalism: The Social Justice Challenge to the Environmental Movement*, ed. Phaedra C. Pezullo and Ronald Sandler (Cambridge, MA: MIT Press, 2007), 105–134; Joanne R. Bauer, *Forging Environmentalism: Justice, Livelihood, and Contested Environments* (Armonk, NY: M. E. Sharpe, 2006), 17–18 and 243.

25. David Schlosberg, *Environmental Justice and the New Pluralism* (New York: Oxford University Press, 1999), 3.

26. Noel Castree characterizes this as a "material essentialism": "A Post-Environmental Ethics?," *Ethics, Place & Environment* 6, no. 1 (March 2003): 3–4, doi:10.1080/13668790303542.

27. For a survey of some, see Ingolfur Blühdorn, *Post-Ecologist Politics: Social Theory and the Abdication of the Ecologist Paradigm* (London: Routledge, 2000), 31–32.

28. Inglehart, "Public Support for Environmental Protection," 65.

29. Inglehart, "Public Support for Environmental Protection."

30. Cf. Dunlap and York, "The Globalization of Environmental Concern and the Limits of the Postmaterialist Values Explanation," 536.

31. Angela G. Mertig and Riley E. Dunlap, "Environmentalism, New Social Movements, and the New Class: A Cross-National Investigation," *Rural Sociology* 66, no. 1 (2001): 113; cf. Juliet Carlisle and Eric R. A. N. Smith, "Postmaterialism vs. Egalitarianism as Predictors of Energy-Related Attitudes," *Environmental Politics* 14, no. 4 (2005): 527–540, doi:10.1080/09644010500215324.

32. Brechin, "Objective Problems, Subjective Values, and Global Environmentalism," 802; for the United States, see Willett M. Kempton, James S. Boster, and Jennifer A. Hartley, *Environmental Values in American Culture* (Cambridge, MA: MIT Press, 1995).

33. Michael Walzer, *Spheres of Justice: A Defense of Pluralism and Equality* (New York: Basic Books, 1983), 8.

34. For a critical yet sympathetic survey, see Catriona Sandilands, *The Good-Natured Feminist: Ecofeminism and the Quest for Democracy* (Minneapolis: University of Minnesota Press, 1999), 3–27.

35. Ariel K. Salleh, *Ecofeminism as Politics: Nature, Marx and the Postmodern* (London: Zed Books, 1997); Mary Mellor, *Feminism & Ecology* (Cambridge, UK: Polity Press, 1997).

36. Sherilyn MacGregor, *Beyond Mothering Earth: Ecological Citizenship and the Politics of Care* (Vancouver: UBC Press, 2011), 71.

37. Ibid.

38. See also Sandilands, *The Good-Natured Feminist*, 48–74.

39. Michael Shellenberger and Ted Nordhaus, "The Death of Environmentalism: Global Warming Politics in a Post-Environmental World," 2004, www.thebreakthrough.org/PDF/Death_of_Environmentalism.pdf.

40. Nordhaus and Shellenberger, *Break Through*, 87.

41. Ibid., 8–9.

42. Ibid., 239.

43. Ibid., 205.

44. Ibid., 22–24; for a more complete discussion of this case, see Jonathan H. Adler, "Fables of the Cuyahoga: Reconstructing a History of Environmental Protection," *Fordham Environmental Law Review* 14 (2002): 89–146.

45. Nordhaus and Shellenberger, *Break Through: From the Death of Environmentalism to the Politics of Possibility*, 28.

46. Ibid., 66–88.

47. Perhaps—although I'm not convinced that even in this "bare life" situation this characterization of materialism would be coherent.

48. E.g., Patrick Novotny, *Where We Live, Work, and Play: The Environmental Justice Movement and the Struggle for a New Environmentalism* (Westport, CT: Praeger, 2000).

49. Hanna Pitkin develops a very similar point in her analysis of Hannah Arendt's dichotomy between "the social" and "the political." See Hanna F. Pitkin, "Justice: On Relating Public and Private," *Political Theory* 9 (1981): 346. I make this point in relation to environmental thought in John M. Meyer, "Review of *The Promise of Green Politics: Environmentalism and the Public Sphere* by Douglas Torgerson," *American Political Science Review* 94, no. 1 (March 2000).

50. Important recent books and edited collections on new materialism include Jane Bennett, *Vibrant Matter: A Political Ecology of Things* (Durham: Duke University Press, 2010); Diana Coole and Samantha Frost, eds., *New Materialisms: Ontology, Agency, and Politics* (Durham: Duke University Press, 2010); Bruce Braun and Sarah Whatmore, eds., *Political Matter: Technoscience, Democracy, and Public Life* (Minneapolis: University of Minnesota Press, 2010); Rick Dolphijn and Iris van der Tuin, *New Materialism: Interviews and Cartographies*, New Metaphysics (London: Open Humanities Press, 2012); Manuel DeLanda, *A New Philosophy of Society: Assemblage Theory and Social Complexity* (London: Continuum International Publishing Group, 2006); Stacy Alaimo, *Bodily Natures: Science, Environment, and the Material Self* (Bloomington: Indiana University Press, 2010); Bruno Latour, *Reassembling the Social: An Introduction to Actor-Network-Theory* (Cambridge: Cambridge University Press, 2007); Stacy Alaimo and Susan Hekman, eds., *Material Feminisms* (Bloomington: Indiana University Press, 2008).

51. Sharon R. Krause, "Bodies in Action: Corporeal Agency and Democratic Politics," *Political Theory* 39, no. 3 (March 2011): 310, doi:10.1177/0090591711400025; Diana Coole and Samantha Frost, "Introducing the New Materialisms," in *New Materialisms: Ontology, Agency, and Politics* (Durham: Duke University Press, 2010), 20.

52. Latour, *Reassembling the Social*, 54–55; Bennett, *Vibrant Matter*, viii–ix, 8–10.

53. Coole and Frost, "Introducing the New Materialisms" offers an especially useful overview of these characteristics.

54. Bennett, *Vibrant Matter*, 4.

55. Ibid., 116–117; in an earlier work, Bennett argues that an "onto-story" can avoid the sorts of problems that I identify here with a focus on ontology. Yet in this later work the distinction is less apparent. Jane Bennett, "The Moraline Drift," in *The Politics of Moralizing*, ed. Jane Bennett and Michael J. Shapiro (New York: Routledge, 2002), 17.

56. Bennett, *Vibrant Matter*, 94.

57. For more on this way of conceptualizing ontology and the prominence of such an "ontological turn" within political theory, see Stephen K. White, *Sustaining Affirmation: The Strengths of Weak Ontology in Political Theory* (Princeton, NJ: Princeton University Press, 2000), 3–17.

58. Bennett, *Vibrant Matter*, 119.

59. Ibid., 111.

60. Two excellent recent papers have raised similar concerns about Bennett's focus on ontology. See Karen Asp, "Trash, Materiality and the Force of Things in Adorno and Bennett" (paper presented at the Western Political Science Association annual meeting, Portland, OR, 2012); Bonnie Washick and Elizabeth Wingrove, "Politics That Matter: Thinking about Power and Justice with New Materialists" (paper presented at the Western Political Science Association annual meeting, Portland, OR, 2012).

61. In the past, I (and many others) have criticized "deep ecology" and other environmental philosophies for this sort of idealism and in particular for their reliance upon a strategy of worldview transformation as a basis for deriving their approach to politics and environmental action: John M. Meyer, *Political Nature: Environmentalism and the Interpretation of Western Thought* (Cambridge, MA: MIT Press, 2001), 22–34 and passim; cf. Ariel K. Salleh, "Deeper than Deep Ecology: The Eco-Feminist Connection," *Environmental Ethics* 6 (Winter 1984): 339–345; Timothy W. Luke, "Deep Ecology as Political Philosophy," in *Eco-Critique: Contesting the Politics of Nature, Economy, and Culture* (Minneapolis: University of Minnesota Press, 1997). Bennett echoes this aim of worldview transformation, especially when she presents politics as derivative of ontology. See her illuminating exchange with Thomas Princen in "Review of *Vibrant Matter: A Political Ecology of Things*. By Jane Bennett," *Perspectives on Politics* 9, no. 1 (2011): 118–120, doi:10.1017/S1537592710003464.

62. Bennett, *Vibrant Matter*, 122.

63. Ibid.; quotation marks in original.

64. Alaimo, *Bodily Natures*, 18.

65. Ibid., 20.

66. Ibid., 113–140 and passim.

67. Teena Gabrielson and Katelyn Parady, "Corporeal Citizenship: Rethinking Green Citizenship through the Body," *Environmental Politics* 19, no. 3 (May 2010): 375, doi:10.1080/09644011003690799.

68. Gay Hawkins, "Plastic Materialities," in *Political Matter: Technoscience, Democracy, and Public Life* (Minneapolis: University of Minnesota Press, 2010), 119.

69. Ibid., 123.

70. Ibid., 125.

71. Ibid., 126–127.

72. Castree, "A Post-Environmental Ethics?," 8.

73. In Di Chiro, "Living Environmentalisms," she discusses these material practices as matters of social reproduction, which also highlights the limitations of the "consumerism" frame. I return to this theme in chapter 7.

74. Bennett has also long expressed concern with this sort of moralizing tone. Unfortunately, her emphasis upon ontological transformation works at cross-purposes with this concern. See *Vibrant Matter*, 12, 38; Bennett, "The Moraline Drift." For a critique of moralism that is targeted to the environmental movement, see William Chaloupka, "The Tragedy of the Ethical Commons: Demoralizing Environmentalism," in *The Politics of Moralizing*, ed. Jane Bennett and Michael J. Shapiro (New York: Routledge, 2002): 113–140.

75. Bruno Latour, "Why Has Critique Run Out of Steam? From Matters of Fact to Matters of Concern," *Critical Inquiry* 30, no. 2 (2004): 231.

76. Ibid., 231–232; cf. Latour, *Reassembling the Social*, 87–120.

4   Private Practices yet Political

1. Jonathan Hafetz, "'A Man's Home Is His Castle?': Reflections on the Home, the Family, and Privacy During the Late Nineteenth and Early Twentieth Centuries," *William & Mary Journal of Women and the Law* 8, no. 2 (February 1, 2002): 180–182.

2. Ibid., 182; cf. Carole Pateman, "Feminist Critiques of the Public/Private Dichotomy," in *The Disorder of Women: Democracy, Feminism, and Political Theory* (Stanford, CA: Stanford University Press, 1989), 118–140; Nancy Fraser, *Unruly Practices: Power, Discourse and Gender in Contemporary Social Theory* (Minneapolis: University of Minnesota Press, 1989), 168–169.

3. Mimi Sheller and John Urry, "Mobile Transformations of 'Public' and 'Private' Life," *Theory, Culture & Society* 20, no. 3 (June 1, 2003): 115, doi:10.1177/02632764030203007.

4. Jeff Weintraub, "The Theory and Politics of the Public/Private Distinction," in *Public and Private in Thought and Practice: Perspectives on a Grand Dichotomy*, ed. Jeff Weintraub and Krishan Kumar (Chicago: University of Chicago Press, 1997), xi.

5. Pateman, "Feminist Critiques of the Public/Private Dichotomy," 120.

6. V. Spike Peterson, "Rereading Public and Private: The Dichotomy That Is Not One," *SAIS Review* 20, no. 2 (2000): 16.

7. John Locke, *Two Treatises of Government*, ed. Peter Laslett (Cambridge: Cambridge University Press, 1988), Second Treatise, chapter V.

8. Raymond Geuss, *Public Goods, Private Goods* (Princeton, NJ: Princeton University Press, 2003), 76–77.

9. Ibid., 79.

10. Ibid., 86; 103.

11. Isaiah Berlin, *Four Essays on Liberty* (Oxford: Oxford University Press, 1969), 122–131.

12. Karl Marx, "On the Jewish Question," in *The Marx-Engels Reader*, ed. Robert C. Tucker (New York: W. W. Norton, 1978).

13. Pateman, "Feminist Critiques of the Public/Private Dichotomy"; Susan Moller Okin, *Justice, Gender and the Family* (New York: Basic Books, 1989).

14. Hannah Arendt, *The Human Condition* (Chicago: University of Chicago Press, 1958), 28–37. There is good reason to doubt the historicity of this as an account of Aristotle and the Greeks, as Bernard Yack has argued convincingly. Bernard Yack, *The Problems of a Political Animal: Community, Justice, and Conflict in Aristotelian Political Thought* (Berkeley: University of California Press, 1993), 9–13; see also my discussion in John M. Meyer, *Political Nature: Environmentalism and the Interpretation of Western Thought* (Cambridge, MA: MIT Press, 2001), 105–108.

15. Arendt, *The Human Condition*, 30–33.

16. Ibid., 37. Arendt's sense of public also includes the constructed world that exists "between" us. Later in the chapter, I seek to make sense of the tensions surrounding material practices in Arendt's public with the assistance of Hanna Pitkin's interpretation of her work.

17. Ibid., 38; 58.

18. Ibid., 28.

19. Ibid., 38.

20. Ibid., 42–43.

21. Ibid., 45.

22. Ibid., 47.

23. Douglas Torgerson, *The Promise of Green Politics: Environmentalism and the Public Sphere* (Durham: Duke University Press, 1999).

24. Ulrich Beck, "Subpolitics: Ecology and the Disintegration of Institutional Power," *Organization & Environment* 10, no. 1 (March 1, 1997): 52–65, doi:10.1177/0921810697101008; Timothy W. Luke, "The People, Politics, and the Planet: Who Knows, Protects, and Serves Nature Best?," in *Democracy and the Claims of Nature*, ed. Ben A. Minteer and Bob Pepperman Taylor (Lanham, MD: Rowman and Littlefield, 2002), 307–308; cf. Noortje Marres, "Front-Staging Nonhumans: Publicity as a Constraint on the Political Activity of Things," in *Political Matter: Technoscience, Democracy, and Public Life*, ed. Bruce Braun and Sarah Whatmore (Minneapolis: University of Minnesota Press, 2010), 177–209.

25. Hanna F. Pitkin, "Justice: On Relating Public and Private," *Political Theory* 9 (1981): 338.

26. Arendt, *The Human Condition*, 52.

27. Ibid., 182.

28. Pitkin, "Justice," 346; cf. Alexandra Kogl, "A Hundred Ways of Beginning: The Politics of Everyday Life," *Polity* 41, no. 4 (2009): 517–523.

29. Cf. Robyn Eckersley, *The Green State: Rethinking Democracy and Sovereignty* (Cambridge, MA: MIT Press, 2004), 140–141; Douglas Torgerson, "Farewell to the Green Movement? Political Action and the Green Public Sphere," *Environmental Politics* 11, no. 1 (2000): 133–145; Catriona Sandilands, *The Good-Natured Feminist: Ecofeminism and the Quest for Democracy* (Minneapolis: University of Minnesota Press, 1999), 159.

30. John Dewey, *Experience and Nature* (George Allen & Unwin Ltd., 1929), 7.

31. Mark B. Brown, *Science in Democracy: Expertise, Institutions, and Representation* (Cambridge, MA: MIT Press, 2009), 144–145.

32. Larry A. Hickman, "Nature as Culture: John Dewey's Pragmatic Naturalism," in *Environmental Pragmatism*, ed. Andrew Light and Erik Katz (London: Routledge, 1996), 54.

33. Martha A. Ackelsberg and Mary Lyndon Shanley, "Privacy, Publicity, and Power: A Feminist Rethinking of the Public-Private Distinction," in *Revisioning the Political: Feminist Reconstructions of Traditional Concepts in Western Political Theory*, ed. Nancy J. Hirschmann and Christine Di Stefano (Boulder, CO: Westview Press, 1996), 222.

34. John Dewey, *The Public and Its Problems* (Athens, OH: Swallow Press, 1927), 20.

35. Hickman, "Nature as Culture."

36. Dewey, *The Public and Its Problems*, 15–16.

37. This point has been developed more recently by Noortje Marres and Mark Brown. cf. Marres, "Front-Staging Nonhumans"; Noortje Marres, "Issues Spark a Public into Being," in *Making Things Public*, ed. Bruno Latour and Peter Weibel (Cambridge, MA: MIT Press, 2005); Brown, *Science in Democracy*, 135–162.

38. Cf. Peter J. Steinberger, "Public and Private," *Political Studies* 47, no. 2 (1999): 309, doi:10.1111/1467-9248.00201.

39. Robert B. Westbrook, *John Dewey and American Democracy* (Ithaca, NY: Cornell University Press, 1993), 460.

40. John Dewey, Jo Ann Boydston, and Lewis S. Feuer, *John Dewey: The Later Works, 1925–1953*, Vol. 15, *1942–1948* (Carbondale: Southern Illinois University Press, 2008), 361.

41. Fred Block, "Introduction," in *The Great Transformation*, by Karl Polanyi (Boston: Beacon Press, 2001), xxxiii.

42. Mitchell Bernard, "Ecology, Political Economy and the Counter-Movement: Karl Polanyi and the Second Great Transformation," in *Innovation and Transformation in International Studies*, ed. Stephen Gill and James H. Mittelman (Cambridge: Cambridge University Press, 1997).

43. Westbrook, *John Dewey and American Democracy*, 314–318.

44. William J. Booth, "On the Idea of the Moral Economy," *American Political Science Review* 88, no. 3 (September 1994): 653.

45. Aristotle, *The Politics* (Chicago: University of Chicago Press, 1984), 1256b35.

46. Ibid., 1257a1.

47. Karl Polanyi, *The Great Transformation: The Political and Economic Origins of Our Time* (Boston: Beacon Press, 1944), 53.

48. Karl Polanyi, *The Livelihood of Man*, ed. Harry W. Pearson (New York: Academic Press, 1977), 19.

49. Ibid.; also Karl Polanyi, *Primitive, Archaic, and Modern: Economies Essays of Karl Polanyi*, ed. George Dalton (Boston: Beacon Press, 1968), 140.

50. Polanyi, *The Livelihood of Man*, 6.

51. Karl Polanyi, "The Economistic Fallacy," in *The Livelihood of Man* (New York: Academic Press, 1977), 5–19.

52. Polanyi, *The Great Transformation*, 3.

53. Polanyi's argument that embeddedness is actually inescapable, despite the best efforts of advocates of market liberalization, is often overlooked or ignored by readers given the tenor of his writing at other points. For example, in an otherwise very careful and systematic essay William James Booth mistakenly asserts that at the core of Polanyi's idea of a "great transformation" is the claim that "in modernity, the economy is disembedded, that is, it is an autonomous sphere governed by laws of its own." Booth, "On the Idea of the Moral Economy," 656. This allows Booth to identify the distinction between an embedded and disembedded economy with other prominent dichotomies in social theory: "premodern" versus "modern" and *Gemeinshaft* versus *Gesellschaft*. Yet Polanyi's argument that a disembedded economy could never exist belies Booth's notion that he views the modern economy as such an autonomous sphere.

54. Bruno Latour, *We Have Never Been Modern* (Cambridge, MA: Harvard University Press, 1993); see also Bruno Latour, *Politics of Nature: How to Bring the Sciences into Democracy* (Cambridge, MA: Harvard University Press, 2004), 10–18, 46.

55. Cf. Block, "Introduction," xxiv.

56. Jack Manno, "Commoditization: Consumption Efficiency and an Economy of Care and Connection," in *Confronting Consumption*, ed. Thomas Princen, Michael Maniates, and Ken Conca (Cambridge, MA: MIT Press, 2002), 77, makes a complementary distinction between goods and services with a "high" or "low" commodity potential. Land, in Manno's terms, would have low commodity potential in part because its use is "embedded . . . in a web of social and ecological relationships."

57. Polanyi, *The Great Transformation*, 72.

58. Ibid., 178.

59. Aldo Leopold, *A Sand County Almanac: And Sketches Here and There* (Oxford: Oxford University Press, 1949), 41.

60. For example, Maria Kaika, "Interrogating the Geographies of the Familiar: Domesticating Nature and Constructing the Autonomy of the Modern Home," *International Journal of Urban and Regional Research* 28, no. 2 (2004): 267, doi:10.1111/j.0309-1317.2004.00519.x; Sarah Whatmore, *Hybrid Geographies: Natures Cultures Spaces* (London: Sage, 2002).

61. Polanyi, *The Great Transformation*, 141.

62. Ibid., 154.

63. Ibid., 162.

64. Ibid., 141.

65. Bernard, "Ecology, Political Economy and the Counter-Movement," 86.

66. Block, "Introduction," xxvii.

67. Manfred B. Steger, *Globalism: The New Market Ideology* (Lanham, MD: Rowman and Littlefield, 2002), 30.

68. Polanyi, *The Great Transformation*, 258a.

69. Ibid., 249–258b.

5 Land and the Concept of Private Property

1. Theodore Steinberg, *Slide Mountain: Or The Folly of Owning Nature* (Berkeley: University of California Press, 1995), 174.

2. See http://www.oregon.gov/LCD/Pages/measure37/legal_information.aspx. More precisely, Oregon's initiative provided cash or waivers only for landowners whose titles predated the regulation being challenged. The peculiar result was that adjacent landowners were granted very different rights depending upon when they initially purchased their property. See Laura Oppenheimer, "Only Thing Developed by Measure 37 Is a Headache," *The Oregonian*, March 6, 2005.

3. Blaine Harden, "Anti-Sprawl Laws, Property Rights Collide in Oregon," *Washington Post*, February 28 2005.

4. Ibid., quoting Harvey Jacobs.

5. Laura Oppenheimer, "Voters Nip Libertarian Dreams across U.S.," *The Oregonian*, November 13, 2006.

6. See Philip Brick and R. McGreggor Cawley, eds., *A Wolf in the Garden: The Land Rights Movement and the New Environmental Debate* (Lanham, MD: Rowman and Littlefield, 1996).

7. Harvey M. Jacobs, "The Future of an American Ideal," in *Private Property in the 21st Century*, ed. Harvey M. Jacobs (Northampton, MA: Edward Elgar, 2004), 173–179.

8. Harden, "Anti-Sprawl Laws, Property Rights Collide in Oregon."

9. Jonathan Walters, "Law of the Land," *Governing Magazine*, May 2005 http://www.governing.com/topics/economic-dev/Law-Land.html.

10. Cf. Michael Walzer, "The Communitarian Critique of Liberalism," *Political Theory* 18, no. 1 (1990): 6–23; Bernard Yack, "Liberalism and Its Communitarian Critics: Does Liberal Practice 'Live Down' to Liberal Theory?" in *Community in America: The Challenges of Habits of the Heart*, ed. Charles H. Reynolds and Ralph V. Norman (Berkeley: University of California Press, 1988), 149–167.

11. By "utopian," I have in mind here its literal definition as *existing nowhere*. See my discussion of Polanyi's contrast between the *utopianism* of laissez-faire economic ideology and the *pragmatism* of resistance to this ideology in chapter 4.

12. Manfred B. Steger, *Globalism: The New Market Ideology* (Lanham, MD: Rowman and Littlefield, 2002).

13. Edward Coke includes this maxim in *The First Part of the Institutes of the Lawes of England*. Steve Sheppard, ed., *The Selected Writings and Speeches of Sir Edward Coke*, vol. II (Indianapolis: Liberty Fund, 2003): 607.

14. Honoré offers one notable exception to this by including a "prohibition of harmful use" as one of his standard incidents of ownership. A. M. Honoré, "Ownership," in *Oxford Essays in Jurisprudence*, ed. A. G. Guest (Oxford: Oxford University Press, 1961), 123. By contrast, Jeremy Waldron argues that this prohibition ought not to be included within the concept of ownership itself—precisely because it is an exception—but ought to instead be treated as a general limit "to what may be done in a given society." Jeremy Waldron, *The Right to Private Property* (Oxford: Clarendon Press, 1988), 32. I critique this quest to purify the concept of property of such notions as "what may be done" in subsequent sections of this chapter.

15. Joseph Singer describes this concept of property as the "ownership model," which he critically contrasts with one that he terms the "entitlement model." Although his analysis of these models is insightful and often echoed here, I find the labels that he applies misleading. Rather than simply equate ownership with the absolutist concept of property, for example, I believe it is more helpful to recognize that the question of what it *means* to "own" property is itself at stake. Cf. Joseph William Singer, *Entitlement: The Paradoxes of Property* (New Haven, CT: Yale University Press, 2000).

16. John Christman, *The Myth of Property: Toward an Egalitarian Theory of Ownership* (New York: Oxford University Press, 1994), 6; Morris R. Cohen, "Property and Sovereignty," *Cornell Law Quarterly* 13 (1927): 11 and passim.

17. John Locke describes the individual in the state of nature with just such imagery: "absolute Lord of his own Person and Possessions, equal to the greatest, and subject to no Body." Yet introducing political sovereignty—the objective of his social contract—necessarily qualifies the right-holder's sovereignty or absolute Lordship. John Locke, *Two Treatises of Government*, ed. Peter Laslett (Cambridge: Cambridge University Press, 1988), sec. 123, 250.

18. Robert Nozick, *Anarchy, State, and Utopia* (New York: Basic Books, 1974), 51–53.

19. We can distinguish this from a view of the political as the location for necessary negotiations between diverse and often competing property claims.

20. Fred Bosselman, David Callies, and John Banta, "The Taking Issue: An Analysis of the Constitutional Limits of Land Use Control" (Washington, DC: Government Printing Office, 1973), 105–123.

21. Cf. Waldron, *The Right to Private Property*, 41, and Cohen, "Property and Sovereignty."

22. Singer, *Entitlement: The Paradoxes of Property*, 4. See also Markku Oksanen, "Environmental Ethics and Concepts of Private Ownership," in *Environmental Ethics and the Global Marketplace*, ed. Dorinda G. Dallmeyer and Albert F. Ike (Athens: University of Georgia Press, 1998), 135.

23. Honoré, "Ownership," 107.

24. Ibid.

25. Waldron, *The Right to Private Property*, 39, 47–53.

26. Bruce A. Ackerman, *Private Property and the Constitution* (New Haven, CT: Yale University Press, 1977), 2.

27. Christman, *The Myth of Property*, 5, makes a similar point.

28. This echoes and develops the earlier analysis of Wesley Newcomb Hohfeld, "Fundamental Legal Conceptions as Applied in Judicial Reasoning," *Yale Law Journal* 26, no. 8 (1917): 710–770; Wesley Newcomb Hohfeld, "Some Fundamental Legal Conceptions as Applied in Judicial Reasoning," *Yale Law Journal* 23, no. 1 (1913): 16–59.

29. Ackerman, *Private Property and the Constitution*, 26–27. Ackerman describes this as the core of what he terms the "scientific policymaker's" understanding of property, which he distinguishes from that of "ordinary observers," who view property in the manner I have described previously.

30. I say "arguably," yet these changes were controversial enough to precipitate legal battles that made their way to the US Supreme Court. Regarding zoning, see *Village of Euclid, Ohio v. Ambler Realty Co.*, 272 US 365 (1926). Regarding airspace, see *US v. Causby*, 328 US 256 (1946).

31. As argued (ultimately unsuccessfully) by the plaintiff in *Penn Central Transport Co. v. New York*, 438 US 104 (1978). For an illuminating discussion of air rights, see "Three-D Deeds: The Rise of Air Rights in New York," in Steinberg, *Slide Mountain*, 135–165.

32. Leif Wenar, "The Concept of Property and the Takings Clause," *Columbia Law Review* 97 (1997): 1930.

33. Richard A. Epstein, *Takings: Private Property and the Power of Eminent Domain* (Cambridge, MA: Harvard University Press, 1985), 57. Wenar, "The Concept of Property and the Takings Clause," 1936.

34. Honoré, "Ownership," 109. Also Alan Ryan, *Property and Political Theory* (Oxford: Basil Blackwell, 1984), 2. Where the question of justification is taken as primary, the question of scope can be ignored, to ill effect. Seana Shiffrin offers a provocative example of this, arguing that theorists seeking to transfer Locke's labor theory of property to a justification for intellectual property typically fail to examine the prior question of whether this form of property would qualify as the sort of thing that Locke would view as necessitating private appropriation from the commons. Seana Valentine Shiffrin, "Lockean Arguments for Private Intellectual Property," in *New Essays in the Legal and Political Theory of Property*, ed. Stephen R. Munzer (Cambridge: Cambridge University Press, 2001), 138–167.

35. David Bollier, *Silent Theft: The Private Plunder of Our Common Wealth* (New York: Routledge, 2003).

36. Karl Polanyi, *The Great Transformation: The Political and Economic Origins of Our Time* (Boston: Beacon Press, 1944), 72; Arild Vatn, "The Environment as a Commodity," *Environmental Values* 9 (2000): 497.

37. Lewis Hyde, *The Gift: Imagination and the Erotic Life of Property* (New York: Random House, 1983).

38. John S Dryzek, *The Politics of the Earth: Environmental Discourses* (Oxford: Oxford University Press, 2005), 75–98.

39. David B. Hunter, "An Ecological Perspective on Property: A Call for Judicial Protection of the Public's Interest in Environmentally Critical Resources," *Harvard Environmental Law Review* 12 (1988): 316–317.

40. Ibid.

41. Joel Kovel, *The Enemy of Nature: The End of Capitalism or the End of the World?* (New York: Zed, 2002), 238–242.

42. Aldo Leopold, *A Sand County Almanac: And Sketches Here and There* (Oxford: Oxford University Press, 1949), 203.

43. Aldo Leopold, "[1947] Forward," in *Companion to a Sand County Almanac*, ed. J. Baird Callicott (Madison: University of Wisconsin Press, 1987), 281.

44. Leopold, *A Sand County Almanac*, 221.

45. Ibid., 224–225. Referring in an unpublished manuscript to "the concept of land as a community"—what he later described as the "B" view—Leopold asserts: "Ninety nine percent of the world's brains and votes have never heard of it. The mass mind is devoid of any notion [of] the integrity of the land community." Quoted in Eric T. Freyfogle, *The Land We Share: Private Property and the Common Good* (Washington, DC: Island Press, 2003), 143.

46. Leopold, *A Sand County Almanac*, 204.

47. Ibid.

48. Garrett Hardin, "The Tragedy of the Commons," *Science* 162 (December 13, 1968). See discussion in Jonathan H. Adler, "Conservative Principles for Environmental Reform" (Case Research Paper Series in Legal Studies, Working Paper 2013-9, March 16, 2013), 109–110, http://papers.ssrn.com/abstract=2234464.

49. On this shift in legal scholarship, see especially Gregory S. Alexander, "Propriety through Commodity? Why Have Legal Environmentalists Embraced Market-Based Solutions?" in *Private Property in the 21st Century: The Future of an American Ideal*, ed. Harvey M. Jacobs (Cheltenham, UK: Edward Elgar, 2004), 75–91; also Carol M. Rose, "The Several Futures of Property: Of Cyberspace and Folk Tales, Emission Trades and Ecosystems," *Minnesota Law Review*, no. 83 (1998): 129–182.

50. Terry L. Anderson and Donald R. Leal, *Free Market Environmentalism*, rev. ed. (New York: Palgrave, 2001), 22; emphasis added. Adler, *Conservative Principles for Environmental Reform*.

51. Tom Bethell, *The Noblest Triumph: Property and Prosperity through the Ages* (New York: St. Martin's Griffin, 1998), 287.

52. Both Mark Sagoff and Markku Oksanen helpfully distinguish between these two forms of appeal to the absolutist concept, though that distinction is not necessary for my argument here. Oksanen, "Environmental Ethics and Concepts of Private Ownership," 121–129; Mark Sagoff, "Free-Market versus Libertarian Environmentalism," *Critical Review* 6, nos. 2–3 (1992): 211–230.

53. Anderson and Leal, *Free Market Environmentalism*, 35, 115, 42.

54. Tibor R. Machan, "Pollution and Political Theory," in *Earthbound: New Introductory Essays in Environmental Ethics*, ed. Tom Regan (New York: Random House, 1984), 97.

55. Sagoff, "Free-Market versus Libertarian Environmentalism," 227.

56. Mark Sagoff, *Price, Principle, and the Environment* (Cambridge: Cambridge University Press, 2004), 116.

57. For example, Anderson and Leal, *Free Market Environmentalism*, 4.

58. Laura S. Underkuffler-Freund, "Property: A Special Right," *Notre Dame Law Review*, no. 71 (1996): 1034.

59. Thomas C. Grey, "The Disintegration of Property," in *NOMOS XXII: Property*, ed. J. Roland Pennock and John W. Chapman (New York: New York University Press, 1980), 69–85.

60. Gary E. Varner, "Environmental Law and the Eclipse of Land as Private Property," in *Ethics and Environmental Policy: Theory Meets Practice*, ed. Frederick Ferre and Peter Hartel (Athens: University of Georgia Press, 1994), 142–160.

61. Grey, "The Disintegration of Property," 69. Grey's contrast here echoes Ackerman's between "scientific policymakers" and "ordinary observers."

62. Ibid., 73.

63. Ibid., 74.

64. Varner, "Environmental Law and the Eclipse of Land as Private Property," 146.

65. Ibid., 144.

66. Ibid., 143.

67. Ibid., 158.

68. MacPherson describes the premodern view as holding property *in* things rather than regarding property as a right *to* things. C. B. MacPherson, "The Meaning of Property," in *Property: Mainstream and Critical Positions*, ed. C. B. MacPherson (Toronto: University of Toronto Press, 1978), 6–9.

69. Louis Hartz, *The Liberal Tradition in America: An Interpretation of American Political Thought since the Revolution* (New York: Harcourt Brace, 1955).

70. William Cronon, *Changes in the Land: Indians, Colonists, and the Ecology of New England* (New York: Hill and Wang, 1983), 75.

71. Ibid., 73.

72. Morton J. Horwitz, *The Transformation of American Law 1780–1860* (Cambridge, MA: Harvard University Press, 1977), 31.

73. Ibid., 36, 31.

74. Freyfogle, *The Land We Share*, 79.

75. Novak quoted in ibid.

76. Horwitz, *The Transformation of American Law 1780–1860*, 40.

77. Gregory S. Alexander, *Commodity and Propriety: Competing Visions of Property in American Legal Thought 1776–1970* (Chicago: University of Chicago Press, 1997), 9, 388n19. Bosselman, Callies, and Banta, "The Taking Issue," 105–123.

78. Bosselman, Callies, and Banta, "The Taking Issue," 319.

79. Freyfogle, *The Land We Share*, 65–84; Horwitz, *The Transformation of American Law 1780–1860*, 31–62.

80. Singer, *Entitlement: The Paradoxes of Property*, 7.

81. For other proposals to reconceptualize property, see Markku Oksanen and Anne-Marie Kumpula, "Is the End of Environmentalism the End of Property? Ownership, the Environment and the Burden of Proof" (paper presented at the ECPR joint sessions: The End of Environmentalism? Turin, 2002); Singer, *Entitlement: The Paradoxes of Property*.

82. Daniel W. Bromley, "Property Rights: Locke, Kant, Peirce and the Logic of Volitional Pragmatism," in *Private Property in the 21st Century: The Future of an American Ideal*, ed. Harvey M. Jacobs (Cheltenham, UK: Edward Elgar, 2004), 26. See also Mark Sagoff, "Takings, Just Compensation, and the Environment," in *Upstream/Downstream: Issues in Environmental Ethics*, ed. Donald Scherer (Philadelphia: Temple University Press, 1990).

83. Ackerman, *Private Property and the Constitution*, 26–27.

84. Singer, *Entitlement: The Paradoxes of Property*, 43.

85. Ibid., 44.

86. See note 29.

87. See note 33.

88. Wenar, "The Concept of Property and the Takings Clause," 1944.

89. Polanyi, *The Great Transformation*, 178.

90. Eric Mortenson, "Voters Approve Land-Use Rules Changes," *The Oregonian*, November 6, 2007.

91. Oregon Department of Land Conservation and Development, *Ballot Measures 37 (2004) and 49 (2007) Outcomes and Effects*, January 2011, 5, http://www.oregon.gov/LCD/docs/publications/m49_2011-01-31.pdf.

92. Eric Mortenson, "Started with Measure 37, Oregon Land-Use War Settled with a Muted Impact on the Land," *The Oregonian*, February 1, 2011, http://www.oregonlive.com/environment/index.ssf/2011/02/oregon_land-use_war_gets_settl.html.

93. Sheila A. Martin et al., "What Is Driving Measure 37 Claims in Oregon?" (paper presented at the Urban Affairs Association annual meeting, Seattle, WA, April 26, 2007), 20–21, http://www.pdx.edu/sites/www.pdx.edu.ims/files/ims_M37April07UAAppt.pdf.

94. Laura Oppenheimer, "Public Demands Land-Use Clarity," *The Oregonian*, February 23, 2007; Laura Oppenheimer, "Oregon, Prepare for a Land-Use Fight, Again," *The Oregonian*, March 30, 2007.

95. In the end, little development actually proceeded under Measure 37 rules. In large part, this is because of the later adoption of Measure 49. However, it also reflects the ambiguities, confusions, and contradictions that resulted from drafting legislation premised upon the supposed simplicity of the absolutist concept of property. See Alex Potapov, "Making Regulatory Takings Reform Work: The Lessons of Oregon's Measure 37," *Environmental Law Reporter News & Analysis* 39 (2009): 10524–10534.

6   Automobility and Freedom

1. Leo Marx, "Technology: The Emergence of a Hazardous Concept," *Social Research* 64, no. 3 (Fall 1997): 981.

2. I frequently use "car," "automobile," and "vehicle" interchangeably. Unless otherwise specified, I use these words colloquially to refer to any form of motor vehicle that is typically individually owned, including trucks, minivans, SUVs, and so on.

3. For an excellent survey of the emergence and uses of "automobility," see Cotten Seiler, *Republic of Drivers: A Cultural History of Automobility in America* (Chicago: University of Chicago Press, 2008), 5–6; Steffen Böhm et al., "Introduction: Impossibilities of Automobility," in *Against Automobility*, edited by Steffen Böhm, Campbell Jones, Chris Land, and Matthew Paterson (Malden, MA: Blackwell, 2006), 4–6.

4. Chella Rajan, one of the few exceptions, argues that "the car and the entire gamut of practices that support it are mistakenly outside the earnest consideration of political theory." Sudhir Chella Rajan, "Automobility, Liberalism, and the Ethics of Driving," *Environmental Ethics* 29 (2007): 77. Cf. Julia Meaton and David Morrice, "The Ethics and Politics of Private Automobile Use,"

Environmental Ethics 18, no. 1 (1996): 39–54; Steve Vanderheiden, "Assessing the Case against the SUV," *Environmental Politics* 15, no. 1 (February 2006): 23–40, doi:10.1080/09644010500418688.

5. Matt DeLorenzo, "Unhorsing the American Cowboy—The Road Ahead: The Automobile as Societal Evil?," *Road and Track*, August 7, 2009, http://www .roadandtrack.com/column/unhorsing-the-american-cowboy.

6. Matthew Paterson, *Automobile Politics: Ecology and Cultural Political Economy* (Cambridge: Cambridge University Press, 2007), 122–123.

7. Paterson (ibid., 91–165) does an especially thorough job of surveying both the political economic and cultural forces at work; for a history of these forces in the United States, see Christopher W. Wells, *Car Country: An Environmental History* (Seattle: University of Washington Press, 2012).

8. Sudhir Chella Rajan, *The Enigma of Automobility: Democratic Politics and Pollution Control* (Pittsburgh: University of Pittsburgh Press, 1996), 33.

9. Seiler, *Republic of Drivers*, 142.

10. Malene Freudendal-Pedersen, *Mobility in Daily Life: Between Freedom and Unfreedom* (London: Ashgate, 2009), 61.

11. Seiler, *Republic of Drivers*, 142.

12. Henry Ford and Samuel Crowther, *My Life and Work* (New York: Doubleday, Page & Company, 1922), 72.

13. Seiler, *Republic of Drivers*, 9.

14. James A. Dunn, "The Politics of Automobility," *The Brookings Review* 17, no. 1 (1999): 43.

15. Freudendal-Pedersen, *Mobility in Daily Life*, 81.

16. Loren E. Lomasky, "Autonomy and Automobility," *Independent Review* 2, no. 1 (1997): 15.

17. Seiler, *Republic of Drivers*, 125.

18. Ibid., 126.

19. Hala Al-Dosari, "Saudi Women Drivers Take the Wheel on June 17—Opinion—Al Jazeera English," *Al Jazeera*, June 16, 2011, http://www.aljazeera .com/indepth/opinion/2011/06/201161694746333674.html; see also http://www .saudiwomendriving.blogspot.de/.

20. TrueCar, "TrueCar.Com Analyzes Vehicle Registration and Gender Differences," *TrueCar* (blog), June 11, 2010, http://blog.truecar.com/2010/06/11/ truecar-com-examines-gender-differences-in-vehicle-registrations/; Gerd Johnsson-Latham, "A Study on Gender Equality as a Prerequisite for Sustainable Development," *Report to the Environment Advisory Council*, 2007, 53, http://www.uft .uni-bremen.de/oekologie/hartmutkoehler_fuer_studierende/MEC/09-MEC -reading/gender%202007%20EAC%20rapport_engelska.pdf.

21. Susan Hanson, "Gender and Mobility: New Approaches for Informing Sustainability," *Gender, Place & Culture* 17, no. 1 (February 2010): 12, doi:10.1080/09663690903498225; Adella Santos et al., *Summary of Travel*

Trends 2009 National Household Travel Survey (US Department of Transportation, June 2011), 29, http://www.osti.gov/scitech/biblio/885762.

22. Doreen Massey quoted in Hanson, "Gender and Mobility," 14. Hanson makes it clear, however, that support for this conclusion is equivocal and context dependent—a point to which I return.

23. Daniel Sperling and Deborah Gordon, *Two Billion Cars: Driving Toward Sustainability* (New York: Oxford University Press, 2010), 13.

24. Ibid.

25. Ibid., 14.

26. Bloomberg News, "China Vehicle Population Hits 240 Million as Smog Engulfs Cities," *Bloomberg*, January 31, 2013, http://www.bloomberg.com/news/2013-02-01/china-vehicle-population-hits-240-million-as-smog-engulfs-cities.html; Sperling and Gordon, *Two Billion Cars*, 210.

27. Lave quoted in Elmer W. Johnson, "Taming the Car and Its User: Should We Do Both?," *Bulletin of the American Academy of Arts and Sciences* 46, no. 2 (November 1, 1992): 14.

28. O'Rourke quoted in Michael C. Moynihan, "Driven Crazy," *Reason*, November 2009, http://reason.com/archives/2009/11/03/driven-crazy.

29. Sperling and Gordon, *Two Billion Cars*, 7.

30. John Urry, "The 'System' of Automobility," *Theory, Culture & Society* 21, nos. 4–5 (October 1, 2004): 28, doi:10.1177/0263276404046059.

31. V. Kaufmann, M. M. Bergman, and D. Joye, "Motility: Mobility as Capital," *International Journal of Urban and Regional Research* 28, no. 4 (2004): 751.

32. See the critical discussion in Katherine J. Goodwin, "Reconstructing Automobility: The Making and Breaking of Modern Transportation," *Global Environmental Politics* 10, no. 4 (November 2010): 70–75.

33. Lomasky, "Autonomy and Automobility," 8.

34. Dunn, "The Politics of Automobility," 40.

35. For the United States, see David Schrank, Bill Eisele, and Tim Lomax, *Urban Mobility Report 2012* (Texas Transportation Institute: Texas A&M University System, December 2012), http://mobility.tamu.edu/ums/.

36. Peter Dauvergne, *The Shadows of Consumption: Consequences for the Global Environment* (Cambridge, MA: MIT Press, 2008), 46.

37. Ibid., 57.

38. Paterson, *Automobile Politics*, 137.

39. Matt Palmquist, "Old without Wheels," *Miller-McCune*, July 14, 2008, http://www.miller-mccune.com/culture-society/old-without-wheels-4419/.

40. L. D. Frank, M. A. Andresen, and T. L. Schmid, "Obesity Relationships with Community Design, Physical Activity, and Time Spent in Cars," *American Journal of Preventive Medicine* 27, no. 2 (2004): 87–96; Howard Frumkin, Lawrence Frank, and Richard J. Jackson, *Urban Sprawl and Public Health: Designing,*

Planning, and Building for Healthy Communities (Washington, DC: Island Press, 2004).

41. Dauvergne, *The Shadows of Consumption*, 56.

42. Ibid., 59–60. Again, this is due to exponential growth in vehicle miles traveled and numbers of vehicles on the road.

43. Paterson provides a more complete survey of these practical challenges: *Automobile Politics*, 32–60.

44. Ibid., 192.

45. Quoted in Rajan, *The Enigma of Automobility*, 23.

46. DeLorenzo, "Unhorsing the American Cowboy."

47. Sperling and Gordon, *Two Billion Cars*, 12.

48. James Q. Wilson, "Cars and Their Enemies," *Commentary*, July 1997, 20.

49. André Gorz, *Ecology as Politics* (Boston: South End Press, 1980), 69, 72.

50. DeLorenzo, "Unhorsing the American Cowboy."

51. Ralph Nader's network of activist organizations, which developed out of his influential challenge to General Motors regarding auto safety, is tellingly named "Public Citizen."

52. Wilson, "Cars and Their Enemies," 21.

53. Rajan, *The Enigma of Automobility*, 8.

54. Paterson, *Automobile Politics*, 221.

55. See Dauvergne, *The Shadows of Consumption*.

56. Urry, "The 'System' of Automobility," 30.

57. Mimi Sheller and John Urry, "Mobile Transformations of 'Public' and 'Private' Life," *Theory, Culture & Society* 20, no. 3 (June 1, 2003): 115, doi:10.1177/02632764030203007.

58. For example: "Public policymakers have a professional predisposition to consider people as so many knights, rooks, and pawns to be moved around on the social chessboard in the service of one's grand strategy. Not all analysts succumb to this temptation, but many do. Their patron saint is the philosopher Plato, the utopian architect of the ideal Republic." Lomasky, "Autonomy and Automobility," 25.

59. "The Ivan Illich Archive—Energy and Equity," http://www.preservenet.com/theory/Illich/EnergyEquity/Energy%20and%20Equity.htm.

60. Adam Millard-Ball and Lee Schipper, "Are We Reaching Peak Travel? Trends in Passenger Transport in Eight Industrialized Countries," *Transport Reviews* 31, no. 3 (May 2011): 372, doi:10.1080/01441647.2010.518291; Santos et al., *Summary of Travel Trends 2009 National Household Travel Survey*.

61. Tony Dutzik and Phineas Baxandall, *A New Direction: Our Changing Relationship with Driving and the Implications for America's Future* (US PIRG Education Fund; Frontier Group, Spring 2013), http://www.uspirg.org/sites/pirg/files/reports/A%20New%20Direction%20vUS.pdf.

62. Elisabeth Rosenthal, "The End of Car Culture," *New York Times*, June 29, 2013, sec. Sunday Review, http://www.nytimes.com/2013/06/30/sunday-review/the-end-of-car-culture.html.

63. Mark Svenvold, "The New Commute," *Orion Magazine*, October 2014, http://www.orionmagazine.org/index.php/articles/article/8269.

64. Dutzik and Baxandall, *A New Direction*; Jeffrey Ball, "The Proportion of Young Americans Who Drive Has Plummeted—And No One Knows Why," *The New Republic*, March 12, 2014, http://www.newrepublic.com/article/116993/millennials-are-abandoning-cars-bikes-carshare-will-it-stick.

65. John Urry, "Automobility, Car Culture and Weightless Travel: A Discussion Paper" (Department of Sociology, Lancaster University, January 1999), 7, http://www.lancs.ac.uk/fass/sociology/papers/urry-automobility.pdf; Lewis Mumford identified this problem of "compulsory mobility" much earlier in *The City in History: Its Origins, Its Transformations, and Its Prospects* (New York: Harcourt Brace Jovanovich, 1961), 503.

66. Santos et al., *Summary of Travel Trends 2009 National Household Travel Survey*; Hanson, "Gender and Mobility," 12–16.

67. Hanson laments the fact that despite the substantial research on the intersection of gender and mobility there has been little done to tease apart this question of choice versus constraint. Hanson, "Gender and Mobility," 15.

68. Urry, "The 'System' of Automobility," 32.

69. Sperling and Gordon, *Two Billion Cars*, 43.

70. For an account of the forces that led to this development in the United States, see Wells, *Car Country*.

71. Freudendal-Pedersen, *Mobility in Daily Life*, 80.

72. Anthony Weston, *Mobilizing the Green Imagination: An Exuberant Manifesto* (Gabriola Island, BC: New Society Publishers, 2012), 49–50.

73. Ball, "The Proportion of Young Americans Who Drive Has Plummeted."

7 Homes, Household Practices, and the Domain(s) of Citizenship

1. William Cronon, "The Trouble with Wilderness; or, Getting Back to the Wrong Nature," in *Uncommon Ground: Toward Reinventing Nature* (New York: W. W. Norton, 1995), 87; 89.

2. Andrew Dobson, *Citizenship and the Environment* (Oxford: Oxford University Press, 2003), 136.

3. Tucker quoted in Robert Gottlieb, *Forcing the Spring: The Transformation of the American Environmental Movement* (Washington, DC: Island Press, 2005), 278.

4. First coined in German by biologist Ernst Haeckel during the nineteenth century as *Oecologie*. Donald Worster, *Nature's Economy: A History of Ecological Ideas* (Cambridge: Cambridge University Press, 1985), 192.

5. M. Nils Peterson, Tarla Rai Peterson, and Jianguo Liu, *The Housing Bomb: Why Our Addiction to Houses Is Destroying the Environment and Threatening Our Society* (Baltimore: Johns Hopkins University Press, 2013), 3–4.

6. Ibid., 15–34.

7. Kersty Hobson, "Environmental Politics, Green Governmentality and the Possibility of a 'Creative Grammar' for Domestic Sustainable Consumption," in *Material Geographies of Household Sustainability*, ed. Ruth Lane and Andrew Gorman-Murray (Farnham: Ashgate, 2011), 202.

8. Julie Reynolds, "Urban Homesteading: The Integral Urban House," *Mother Earth News*, November/December 1976, http://www.motherearthnews.com/modern-homesteading/urban-homesteading-zmaz76ndztak.aspx.

9. E. F. Schumacher, *Small Is Beautiful: Economics as if People Mattered* (New York: Harper and Row, 1973).

10. Langdon Winner, *The Whale and the Reactor: A Search for Limits in an Age of High Technology* (Chicago: University of Chicago Press, 1986), 80; for a more recent critique of technological exuberance as a theory of environmental change, see Damian F. White, "A Green Industrial Revolution? Sustainable Technological Innovation in a Global Age," *Environmental Politics* 11, no. 2 (June 2002): 1–26, doi:10.1080/714000603.

11. "About CCAT," *Campus Center for Appropriate Technology*, n.d., http://www.ccathsu.com/about.

12. Martin Heidegger, "Building, Dwelling, Thinking," in *Basic Writings*, rev. ed., ed. David Farrell Krell (HarperCollins, 1993), 348.

13. Iris Marion Young, "House and Home: Feminist Variations on a Theme," in *Feminist Interpretations of Martin Heidegger* (University Park: Pennsylvania State University Press, 2001), 255–256.

14. Ibid., 269; for a more complete development of the relationship of building or "founding" to preservation, see Peter F. Cannavò, *The Working Landscape: Founding, Preservation, and the Politics of Place* (Cambridge, MA: MIT Press, 2007).

15. Sean Armstrong, "Notes for CCAT History," n.d., unpublished manuscript.

16. Ashley Katz, "About LEED," *US Green Building Council*, July 25, 2012, http://www.usgbc.org/articles/about-leed; Noortje Marres, "Testing Powers of Engagement: Green Living Experiments, the Ontological Turn and the Undoability of Involvement," *European Journal of Social Theory* 12, no. 1 (February 1, 2009): 117–133, doi:10.1177/1368431008099647; Noortje Marres, *Material Participation: Technology, the Environment and Everyday Publics* (London: Palgrave Macmillan, 2012), 106–131.

17. Marres, *Material Participation*, 114.

18. Bruno Latour, "Why Has Critique Run out of Steam? From Matters of Fact to Matters of Concern," *Critical Inquiry* 30, no. 2 (2004): 225–248.

19. For example, Kevin Fagan, "Power on a Shoestring/21-Year Experiment Finds Ways Anyone Can Cut Energy Costs," *San Francisco Chronicle*, January 8, 2001.

20. Alison Blunt and Robyn Dowling, *Home* (London: Routledge, 2006), 2, 22; Peter Saunders and Peter Williams, "The Constitution of the Home: Towards a Research Agenda," *Housing Studies* 3, no. 2 (1988): 81–93, doi:10.1080/02673038808720618.

21. Linda A. Jacobsen, Mark Mather, and Genevieve Dupuis, "Household Change in the United States," *Population Reference Bureau*, September 2012, http://www.prb.org/Publications/PopulationBulletins/2012/us-household -change.aspx.

22. Winner, *The Whale and the Reactor*, 64.

23. Ibid., 79.

24. Gottlieb, *Forcing the Spring*, 279.

25. Mark Sagoff, *The Economy of the Earth: Philosophy, Law and the Environment* (Cambridge: Cambridge University Press, 1988), 50–51.

26. Ibid., 52.

27. Ibid., 53.

28. Ibid., 52.

29. Russell Keat raised a related challenge to Sagoff's dichotomy two decades ago, arguing that the pursuit of consumption is often a publicly shared value and not merely a private preference: "Citizens, Consumers and the Environment: Reflections on *The Economy of the Earth*," *Environmental Values* 3, no. 4 (November 1, 1994): 333–349, doi:10.3197/096327194776679674.

30. Michael Maniates, "Individualization Plant a Tree, Buy a Bike, Save the World?," in *Confronting Consumption*, ed. Thomas Princen, Michael Maniates, and Ken Conca (Cambridge, MA: MIT Press, 2002).

31. See also Andrew Szasz, *Shopping Our Way to Safety: How We Changed from Protecting the Environment to Protecting Ourselves* (Minneapolis: University of Minnesota Press, 2009); Timothy W. Luke, *Ecocritique: Contesting the Politics of Nature, Economy, and Culture* (Minneapolis: University of Minnesota Press, 1997), 116–136.

32. Dobson, *Citizenship and the Environment*, 83–139; see further discussion of Dobson's argument in chapter 2.

33. Kersty Hobson, "On the Making of the Environmental Citizen," *Environmental Politics* 22, no. 1 (February 2013): 58, doi:10.1080/09644016.2013.755 388.

34. Sherilyn MacGregor, "Ecological Citizenship," in *Handbook of Political Citizenship and Social Movements*, ed. H.-A. van der Heijden (Cheltenham, UK: Edward Elgar, 2014), 118.

35. Rasmus Karlsson, "Individual Guilt or Collective Progressive Action? Challenging the Strategic Potential of Environmental Citizenship Theory," *Environmental Values* 21, no. 4 (November 1, 2012): 460, doi:10.3197/0963271 12X13466893628102; see also Sherilyn MacGregor, *Beyond Mothering Earth: Ecological Citizenship and the Politics of Care* (Vancouver: UBC Press, 2011), 105–112.

36. Dolores Hayden, *Redesigning the American Dream: Gender, Housing, and Family Life*, rev. ed. (New York: W. W. Norton, 2002), 87.

37. Ibid., 88.

38. Robert Bruegmann, *Sprawl: A Compact History* (University of Chicago Press, 2006); Mark Mather, Kevin Pollard, and Linda A. Jacobsen, *Reports America* (Washington, DC: Population Reference Bureau, 2009), 1–15; Fiona Allon, *Renovation Nation: Our Obsession with Home* (Sydney, Australia: UNSW Press, 2008), 65–70.

39. In addition to previous sources, see J. H. Kunstler, *The Geography of Nowhere: The Rise and Decline of America's Man-Made Landscape* (New York: Simon and Schuster, 1993).

40. Anthony Flint, *This Land: The Battle over Sprawl and the Future of America* (Baltimore: Johns Hopkins University Press, 2008), 49–58. Another strand of criticism, especially in the United States, has focused on the economic and racial exclusivity of many suburbs. Yet here the picture has changed substantially in recent years, with US suburbs, overall, now having levels of both poverty and racial diversity as high as or higher than cities; see Myron Orfield and Thomas Luce, "America's Racially Diverse Suburbs: Opportunities and Challenges," *University of Minnesota Law School*, July 20, 2012, http://www.law.umn.edu/uploads/e0/65/e065d82a1c1da0bfef7d86172ec5391e/Diverse_Suburbs_FINAL.pdf.

41. Cannavò, *The Working Landscape*, 93–122.

42. Flint, *This Land*, 12.

43. Allon, *Renovation Nation*.

44. Cannavò, *The Working Landscape*, 97–98; Joel Kotkin, "The Triumph of Suburbia," *Newgeography.com*, April 29, 2013, http://www.newgeography.com/content/003667-the-triumph-suburbia.

45. Richard Flacks made this point in a somewhat different context: "Thus the fully equipped suburban home provides a framework for living that fundamentally threatens the possibility of public life . . . But it is crucial to understand it also as a space whose attractions include a sense of safety and security and also a sense of control and choice." *Making History: The American Left and the American Mind* (New York: Columbia University Press, 1988), 43.

46. Chris Gibson et al., "Is It Easy Being Green? On the Dilemmas of Material Cultures of Household Sustainability," in *Material Geographies of Household Sustainability*, ed. Ruth Lane and Andrew Gorman-Murray (Farnham: Ashgate, 2011), 19–34.

47. Teena Gabrielson and Katelyn Parady, "Corporeal Citizenship: Rethinking Green Citizenship through the Body," *Environmental Politics* 19, no. 3 (May 2010): 383, doi:10.1080/09644011003690799.

48. Fiona Allon, "Ethical Consumption Begins at Home: Green Renovations, Eco-Homes and Sustainable Home Improvement," in *Ethical Consumption: A Critical Introduction*, ed. Tania Lewis and Emily Potter (Abingdon, Oxon: Routledge, 2011), 207.

49. David Harvey, "The Right to the City," *New Left Review* 53 (October 2008): 27.

50. For example, Susan Moller Okin, *Justice, Gender and the Family* (New York: Basic Books, 1989), 22–23, 131–133.

51. Thad Williamson, *Sprawl, Justice, and Citizenship: The Civic Costs of the American Way of Life* (New York: Oxford University Press, 2010), 4.

52. Although as Hayden's account of these nineteenth-century strategies makes clear, they rarely sought to redistribute this responsibility away from women as a whole or to challenge the gender identification of such care: *Redesigning the American Dream*, 91–95.

53. Ibid., 177–181; Graham Meltzer, *Sustainable Community: Learning from the Cohousing Model* (Bloomington, IN: Trafford Publishing, 2005); Louise Crabtree, "Disintegrated Houses: Exploring Ecofeminist Housing and Urban Design Options," *Antipode* 38, no. 4 (2006): 711–734, doi:10.1111/j.1467-8330.2006.00473.x.

54. Hayden, *Redesigning the American Dream*, 204–211.

55. Tina Rosenberg, "It's Not Just Nice to Share, It's the Future," *New York Times*, June 5, 2013, http://opinionator.blogs.nytimes.com/2013/06/05/its-not-just-nice-to-share-its-the-future/; Timothy D. May, "Shared Housing: The Sharing Economy Gives Roommates a New Image," *Christian Science Monitor*, March 30, 2014, http://www.csmonitor.com/The-Culture/Family/2014/0330/Shared-housing-The-sharing-economy-gives-roommates-a-new-image.

56. Julian Agyeman, Duncan McLaren, and Adrianne Schaefer-Borrego, "Sharing Cities," *Friends of the Earth*, "Big Ideas" Project, September 2013, http://www.foe.co.uk/sites/default/files/downloads/agyeman_sharing_cities.pdf; Adam Parsons, "The Sharing Economy: A Short Introduction to Its Political Evolution," *openDemocracy*, March 5, 2014, http://www.opendemocracy.net/transformation/adam-parsons/sharing-economy-short-introduction-to-its-political-evolution.

57. Robert Kuttner, "Share Economy or Bare Economy?," *Huffington Post*, April 27, 2014, http://www.huffingtonpost.com/robert-kuttner/share-economy-or-bare-eco_b_5223968.html.

58. MacGregor, *Beyond Mothering Earth*, 184.

59. Vanessa Taylor and Frank Trentmann, "Liquid Politics: Water and the Politics of Everyday Life in the Modern City," *Past & Present* 211, no. 1 (May 1, 2011): 199–241, doi:10.1093/pastj/gtq068.

60. Ibid., 203–204.

61. Hobson, "'Creative Grammar' for Domestic Sustainable Consumption," 198–206; Marres, *Material Participation*.

62. Hanna F. Pitkin, "Justice: On Relating Public and Private," *Political Theory* 9 (1981): 346. See my discussion of this point in chapter 4.

63. Witold Rybczynski, *Home: A Short History of an Idea* (Harmondsworth: Penguin Books, 1987), 21–22.

64. Ibid., 38–39.

65. Elizabeth Shove, *Comfort, Cleanliness and Convenience: The Social Organization of Normality* (Oxford, UK: Berg, 2003), 198.

66. Ibid., 21–77.

67. Ibid., 141–151.

68. Ibid., 199.

69. Margaret M. Willis and Juliet B. Schor, "Does Changing a Light Bulb Lead to Changing the World? Political Action and the Conscious Consumer," *The Annals of the American Academy of Political and Social Science* 644, no. 1 (November 1, 2012): 160–190, doi:10.1177/0002716212454831; Lucy Atkinson, "Buying In to Social Change: How Private Consumption Choices Engender Concern for the Collective," *The Annals of the American Academy of Political and Social Science* 644, no. 1 (November 1, 2012): 191–206, doi:10.1177/0002716212448366.

70. Maria Kaika, "Interrogating the Geographies of the Familiar: Domesticating Nature and Constructing the Autonomy of the Modern Home," *International Journal of Urban and Regional Research* 28, no. 2 (2004): 266, doi:10.1111/j.0309-1317.2004.00519.x.

71. Ibid.

72. See, including footnotes, Taylor and Trentmann, "Liquid Politics," 200–201 and passim.

73. Kaika, "Interrogating the Geographies of the Familiar," 274.

74. Ibid., 275.

75. Ibid., 276.

76. Ibid., 283.

77. Giovanna Di Chiro, "Living Environmentalisms: Coalition Politics, Social Reproduction, and Environmental Justice," *Environmental Politics* 17, no. 2 (2008): 285–287, doi:10.1080/09644010801936230; see also Isabella Bakker, "Social Reproduction and the Constitution of a Gendered Political Economy," *New Political Economy* 12, no. 4 (2007): 541–556, doi:10.1080/13563460701661561.

8   Conclusion

1. William Ophuls, *Ecology and the Politics of Scarcity: Prologue to a Political Theory of the Steady State* (San Francisco: W. H. Freeman, 1977); Robert Heilbroner, *An Inquiry into the Human Prospect: Updated and Reconsidered for the 1980s* (New York: Norton, 1980); Garrett Hardin, "The Tragedy of the Commons," *Science* 162 (December 13, 1968): 1243–1248.

2. John Barry, *The Politics of Actually Existing Unsustainability* (Oxford: Oxford University Press, 2012), 20–22.

3. James E. Lovelock, *The Vanishing Face of Gaia: The Final Warning* (New York: Basic Books, 2009); David J. C. Shearman and Joseph Wayne Smith, *The Climate Change Challenge and the Failure of Democracy* (Westport, CT: Praeger Publishers, 2007); for a recent critique, see Pascal Bruckner, *The Fanaticism of the Apocalypse: Save the Earth, Punish Human Beings* (Cambridge, UK: Polity Press, 2013).

4. Paul Hawken, Amory Lovins and Hunter Lovins, *Natural Capitalism: Creating the Next Industrial Revolution* (Boston: Little, Brown and Company, 1999); William McDonough and Michael Braungart, *Cradle to Cradle: Remaking the Way We Make Things* (New York: North Point Press, 2002); see also Rocky Mountain Institute, "Reinventing Fire," n.d., http://www.rmi.org/reinventingfire.

5. Hawken et al., *Natural Capitalism*, 1–2; emphasis added.

6. The classic source for a discussion of "autonomous" technological development as the displacement of politics is Langdon Winner, *Autonomous Technology: Technics-out-of-Control as a Theme in Political Thought* (Cambridge, MA: MIT Press, 1977).

7. Hawken et al., *Natural Capitalism*, 20.

8. Fred Pearce, "New Green Vision: Technology as Our Planet's Last Best Hope," *Yale Environment 360*, July 15, 2013, http://e360.yale.edu/feature/new_green_vision_technology_as_our_planets_last_best_hope/2671/.

9. Rasmus Karlsson, "Ambivalence, Irony, and Democracy in the Anthropocene," *Futures* 46 (February 2013): 1, doi:10.1016/j.futures.2012.12.002.

10. Ted Nordhaus and Michael Shellenberger, *Break Through: From the Death of Environmentalism to the Politics of Possibility* (Boston: Houghton Mifflin, 2007).

11. For further development of this theme, see John M. Meyer, "A Democratic Politics of Sacrifice?," in *The Environmental Politics of Sacrifice*, ed. Michael Maniates and John M. Meyer (Cambridge, MA: MIT Press, 2010), 13–32; Jonathan Lear, *Radical Hope: Ethics in the Face of Cultural Devastation* (Cambridge, MA: Harvard University Press, 2006); Rebecca Solnit, "Hope: The Care and Feeding of," *Grist*, August 2, 2011, http://grist.org/living/2011-08-02-hope-the-care-and-feeding-of/; Allen Thompson, "Radical Hope for Living Well in a Warmer World," *Journal of Agricultural and Environmental Ethics* 23, no. 1–2 (June 17, 2009): 43–59, doi:10.1007/s10806-009-9185-2; Barry, *The Politics of Actually Existing Unsustainability*, 54–61.

12. Jean-Jacques Rousseau, *Discourse on the Origin of Inequality*, trans. Donald A. Cress (Indianapolis: Hackett Publishing, 1992), 48.

Bibliography

Abbey, Ruth. "Is Liberalism Now an Essentially Contested Concept?" *New Political Science* 27, no. 4 (December 2005): 461–480. doi:10.1080/07393140500370972.

"About CCAT." *Campus Center for Appropriate Technology*, n.d. Accessed March 8, 2013. http://www.ccathsu.com/about.

Ackelsberg, Martha A., and Mary Lyndon Shanley. "Privacy, Publicity, and Power: A Feminist Rethinking of the Public-Private Distinction." In *Revisioning the Political: Feminist Reconstructions of Traditional Concepts in Western Political Theory*, edited by Nancy J. Hirschmann and Christine Di Stefano, 213–233. Boulder, CO: Westview Press, 1996.

Ackerman, Bruce A. *Private Property and the Constitution.* New Haven, CT: Yale University Press, 1977.

Adler, Jonathan H. "Conservative Principles for Environmental Reform." Case Research Paper Series in Legal Studies, Working Paper 2013-9, March 16, 2013. Accessed June 3, 2013. http://papers.ssrn.com/abstract=2234464.

Adler, Jonathan H. "Fables of the Cuyahoga: Reconstructing a History of Environmental Protection." *Fordham Environmental Law Review* 14 (2002): 89–146.

Agyeman, Julian. *Introducing Just Sustainabilities: Policy, Planning and Practice.* London: Zed Books, 2013.

Agyeman, Julian, McLaren, Duncan, and Schaefer-Borrego, Adrianne. "Sharing Cities." *Friends of the Earth*, "Big Ideas" Project, September 2013. Accessed April 1, 2014. http://www.foe.co.uk/sites/default/files/downloads/agyeman_sharing_cities.pdf.

Alaimo, Stacy. *Bodily Natures: Science, Environment, and the Material Self.* Bloomington: Indiana University Press, 2010.

Alaimo, Stacy, and Susan Hekman, eds. *Material Feminisms.* Bloomington: Indiana University Press, 2008.

Al-Dosari, Hala. "Saudi Women Drivers Take the Wheel on June 17." *Al Jazeera*, June 16, 2011. Accessed July 29, 2011. http://www.aljazeera.com/indepth/opinion/2011/06/201161694746333674.html.

Alexander, Gregory S. *Commodity and Propriety: Competing Visions of Property in American Legal Thought 1776–1970*. Chicago: University of Chicago Press, 1997.

Alexander, Gregory S. "Propriety through Commodity? Why Have Legal Environmentalists Embraced Market-Based Solutions?" In *Private Property in the 21st Century: The Future of an American Ideal*, edited by Harvey M. Jacobs, 75–91. Cheltenham, UK: Edward Elgar, 2004.

Allen, Kim, Vinci Daro, and Dorothy Holland. "Becoming an Environmental Justice Activist." In *Environmental Justice and Environmentalism: The Social Justice Challenge to the Environmental Movement*, edited by Phaedra C. Pezullo and Ronald Sandler, 105–134. Cambridge, MA: MIT Press, 2007.

Allon, Fiona. "Ethical Consumption Begins at Home: Green Renovations, Eco-Homes and Sustainable Home Improvement." In *Ethical Consumption: A Critical Introduction*, edited by Tania Lewis and Emily Potter, 202–215. Abingdon, Oxon: Routledge, 2010.

Allon, Fiona. *Renovation Nation: Our Obsession with Home*. Sydney, Australia: UNSW Press, 2008.

Anderson, Terry L., and Donald R. Leal. *Free Market Environmentalism*. Rev. ed. New York: Palgrave, 2001.

Arendt, Hannah. *The Human Condition*. Chicago: University of Chicago Press, 1958.

Arias-Maldonado, Manuel. *Real Green: Sustainability after the End of Nature*. Farnham: Ashgate, 2012.

Aristotle. *The Politics*. Chicago: University of Chicago Press, 1984.

Armstrong, Sean. "Notes for CCAT History," n.d. Unpublished manuscript.

Asp, Karen. "Trash, Materiality and the Force of Things in Adorno and Bennett." Paper presented at the Western Political Science Association annual meeting, Portland, OR, 2012.

Atkinson, Lucy. "Buying In to Social Change: How Private Consumption Choices Engender Concern for the Collective." *The Annals of the American Academy of Political and Social Science* 644, no. 1 (November 1, 2012): 191–206. doi:10.1177/0002716212448366.

Bakker, Isabella. "Social Reproduction and the Constitution of a Gendered Political Economy." *New Political Economy* 12, no. 4 (2007): 541–556. doi: 10.1080/13563460701661561.

Ball, Jeffrey. "The Proportion of Young Americans Who Drive Has Plummeted—And No One Knows Why." *The New Republic*, March 12, 2014. Accessed March 20, 2014. http://www.newrepublic.com/article/116993/millennials-are -abandoning-cars-bikes-carshare-will-it-stick.

Barry, Brian. "Sustainability and Intergenerational Justice." In *Fairness and Futurity: Essays on Environmental Sustainability and Social Justice*, edited by Andrew Dobson, 93–117. Oxford: Oxford University Press, 1999.

Barry, John. "Greening Liberal Democracy: Practice, Theory and Political Economy." In *Sustaining Liberal Democracy*, edited by John Barry and Marcel Wissenburg, 59–80. New York: Palgrave, 2001.

Barry, John. *The Politics of Actually Existing Unsustainability*. Oxford: Oxford University Press, 2012.

Barry, John, and Marcel Wissenburg. *Sustaining Liberal Democracy: Ecological Challenges and Opportunities*. New York: Palgrave, 2001.

Bauer, Joanne R. *Forging Environmentalism: Justice, Livelihood, and Contested Environments*. Armonk, NY: M. E. Sharpe, 2006.

Beck, Ulrich. "Subpolitics: Ecology and the Disintegration of Institutional Power." *Organization & Environment* 10, no. 1 (March 1, 1997): 52–65. doi:10.1177/0921810697101008.

Bell, Derek. "How Can Political Liberals Be Environmentalists?" *Political Studies* 50, no. 4 (2002): 703–724. doi:10.1111/1467-9248.00003.

Bennett, Jane. "The Moraline Drift." In *The Politics of Moralizing*, edited by Jane Bennett and Michael J. Shapiro, 11–26. New York: Routledge, 2002.

Bennett, Jane. *Vibrant Matter: A Political Ecology of Things*. Durham: Duke University Press, 2010.

Berlin, Isaiah. *Four Essays on Liberty*. Oxford: Oxford University Press, 1969.

Bernard, Mitchell. "Ecology, Political Economy and the Counter-Movement: Karl Polanyi and the Second Great Transformation." In *Innovation and Transformation in International Studies*, edited by Stephen Gill and James H. Mittelman, 75–89. Cambridge: Cambridge University Press, 1997.

Bernstein, Eduard. *Evolutionary Socialism*. New York: B. W. Huebsch, 1911.

Bethell, Tom. *The Noblest Triumph: Property and Prosperity through the Ages*. New York: St. Martin's Griffin, 1998.

"Blast from the Past - Values Party TV Ad," June 1, 2012. Accessed December 18, 2012. http://www.3news.co.nz/Blast-from-the-past---Values-Party-TV-ad/tabid/315/articleID/256413/Default.aspx.

Block, Fred. "Introduction." In *The Great Transformation, by Karl Polanyi*, xviii–xxxviii. Boston: Beacon Press, 2001.

Bloomberg News. "China Vehicle Population Hits 240 Million as Smog Engulfs Cities." *Bloomberg*, January 31, 2013. Accessed July 26, 2013. http://www.bloomberg.com/news/2013-02-01/china-vehicle-population-hits-240-million-as-smog-engulfs-cities.html.

Blühdorn, Ingolfur. "The Governance of Unsustainability: Ecology and Democracy after the Post-Democratic Turn." *Environmental Politics* 22, no. 1 (February 2013): 16–36. doi:10.1080/09644016.2013.755005.

Blühdorn, Ingolfur. "The Politics of Unsustainability: COP15, Post-Ecologism, and the Ecological Paradox." *Organization & Environment* 24, no. 1 (March 1, 2011): 34–53. doi:10.1177/1086026611402008.

Blühdorn, Ingolfur. *Post-Ecologist Politics: Social Theory and the Abdication of the Ecologist Paradigm*. London: Routledge, 2000.

Blunt, Alison, and Robyn Dowling. *Home*. London: Routledge, 2006.

Böhm, Steffen, Campbell Jones, Chris Land, and Mat Paterson. "Introduction: Impossibilities of Automobility." In *Against Automobility*, edited by Steffen Böhm, Campbell Jones, Chris Land, and Matthew Paterson, 1–16. Malden, MA: Blackwell, 2006.

Bollier, David. *Silent Theft: The Private Plunder of Our Common Wealth*. New York: Routledge, 2003.

Booth, William J. "On the Idea of the Moral Economy." *American Political Science Review* 88, no. 3 (September 1994): 653-667.

Bosselman, Fred, David Callies, and John Banta. "The Taking Issue: An Analysis of the Constitutional Limits of Land Use Control." Washington, DC: Government Printing Office, 1973.

Braun, Bruce, and Sarah Whatmore, eds. *Political Matter: Technoscience, Democracy, and Public Life*. Minneapolis: University of Minnesota Press, 2010.

Brechin, Steven R. "Objective Problems, Subjective Values, and Global Environmentalism: Evaluating the Postmaterialist Argument and Challenging a New Explanation." *Social Science Quarterly* 80, no. 4 (December 1999): 793–809.

Brick, Philip, and R. McGreggor Cawley, eds. *A Wolf in the Garden: The Land Rights Movement and the New Environmental Debate*. Lanham, MD: Rowman and Littlefield, 1996.

Bromley, Daniel W. "Property Rights: Locke, Kant, Peirce and the Logic of Volitional Pragmatism." In *Private Property in the 21st Century: The Future of an American Ideal*, edited by Harvey M. Jacobs, 19–30. Cheltenham, UK: Edward Elgar, 2004.

Brown, Mark B. *Science in Democracy: Expertise, Institutions, and Representation*. Cambridge, MA: MIT Press, 2009.

Bruckner, Pascal. *The Fanaticism of the Apocalypse: Save the Earth, Punish Human Beings*. Cambridge, UK: Polity Press, 2013.

Bruegmann, Robert. *Sprawl: A Compact History*. Chicago: University of Chicago Press, 2006.

Buck, Christopher D. "Post-Environmentalism: An Internal Critique." *Environmental Politics* 22, no. 6 (November 2013): 883–900. doi:10.1080/09644016.2012.712793.

Buford, Talia. "Greens Confront Own Need for Diversity." *POLITICO*, December 29, 2012. Accessed April 15, 2013. http://www.politico.com/story/2012/12/greens-confront-own-need-for-diversity-85558.html.

Calder, Gideon, and Catriona McKinnon. "Introduction: Climate Change and Liberal Priorities." *Critical Review of International Social and Political Philosophy* 14, no. 2 (2011): 91–97. doi:10.1080/13698230.2011.529702.

Cannavò, Peter F. *The Working Landscape: Founding, Preservation, and the Politics of Place*. Cambridge, MA: MIT Press, 2007.

Carlisle, Juliet, and Eric R. A. N. Smith. "Postmaterialism vs. Egalitarianism as Predictors of Energy-Related Attitudes." *Environmental Politics* 14, no. 4 (August 2005): 527–540. doi:10.1080/09644010500215324.

Castree, Noel. "A Post-Environmental Ethics?" *Ethics, Place & Environment* 6, no. 1 (March 2003): 3–12. doi:10.1080/13668790303542.

Chaloupka, William. "The Tragedy of the Ethical Commons: Demoralizing Environmentalism." In *The Politics of Moralizing*, edited by Jane Bennett and Michael J. Shapiro, 113–140. New York: Routledge, 2002.

Christman, John. *The Myth of Property: Toward an Egalitarian Theory of Ownership*. New York: Oxford University Press, 1994.

Cohen, Morris R. "Property and Sovereignty." *Cornell Law Quarterly* 13 (1927): 8–30.

Connolly, William E. "The Evangelical-Capitalist Resonance Machine." *Political Theory* 33, no. 6 (December 2005): 869–886. doi:10.1177/0090591705280376.

Coole, Diana, and Samantha Frost, eds. *New Materialisms: Ontology, Agency, and Politics*. Durham: Duke University Press, 2010.

Coole, Diana, and Samantha Frost. "Introducing the New Materialisms." In *New Materialisms: Ontology, Agency, and Politics*, 1–46. Durham: Duke University Press, 2010.

Crabtree, Louise. "Disintegrated Houses: Exploring Ecofeminist Housing and Urban Design Options." *Antipode* 38, no. 4 (2006): 711–734. doi:10.1111/j.1467-8330.2006.00473.x.

Cronon, William. *Changes in the Land: Indians, Colonists, and the Ecology of New England*. New York: Hill and Wang, 1983.

Cronon, William. "The Trouble with Wilderness; or, Getting Back to the Wrong Nature." In *Uncommon Ground: Toward Reinventing Nature*, 69–90. New York: W. W. Norton, 1995.

Dagger, Richard. "Freedom and Rights." In *Political Theory and the Ecological Challenge*, edited by Andrew Dobson and Robyn Eckersley, 200–215. Cambridge: Cambridge University Press, 2006.

Dann, Christine. "'From Earth's Last Islands': The Development of the First Two Green Parties New Zealand and Tasmania," n.d. Accessed August 19, 2014. http://www.globalgreens.org/earths-last-islands-development-first-two-green-parties-tasmania-and-new-zealand.

Dauvergne, Peter. *The Shadows of Consumption: Consequences for the Global Environment*. Cambridge, MA: MIT Press, 2008.

DeLanda, Manuel. *A New Philosophy of Society: Assemblage Theory and Social Complexity*. London: Continuum International Publishing Group, 2006.

DeLorenzo, Matt. "Unhorsing the American Cowboy—The Road Ahead: The Automobile as Societal Evil?" *Road and Track*, August 7, 2009. Accessed

August 19, 2014. http://www.roadandtrack.com/column/unhorsing-the-american -cowboy.

de-Shalit, Avner. *The Environment between Theory and Practice*. Oxford: Oxford University Press, 2000.

Dewey, John, Jo Ann Boydston, and Lewis S. Feuer. *John Dewey: The Later Works, 1925–1953*. Vol. 15, *1942–1948*. Carbondale: Southern Illinois University Press, 2008.

Dewey, John. *Experience and Nature*. London: George Allen & Unwin Ltd., 1929.

Dewey, John. *The Public and Its Problems*. Athens, OH: Swallow Press, 1927.

Di Chiro, Giovanna. "Living Environmentalisms: Coalition Politics, Social Reproduction, and Environmental Justice." *Environmental Politics* 17, no. 2 (2008): 276–298. doi:10.1080/09644010801936230.

diZerega, Gus. "Empathy, Society, Nature and the Relational Self: Deep Ecology and Liberal Modernity." *Social Theory and Practice* 21 (Summer 1995): 239–269.

Dobson, Andrew. *Citizenship and the Environment*. Oxford: Oxford University Press, 2003.

Dobson, Andrew. *Green Political Thought*. Second ed. London: Routledge, 1995.

Dobson, Andrew. *Justice and the Environment: Conceptions of Environmental Sustainability and Dimensions of Social Justice*. Oxford: Oxford University Press, 1998.

Dolphijn, Rick, and Iris van der Tuin. *New Materialism: Interviews and Cartographies*. New Metaphysics series. London: Open Humanities Press, 2012.

Dryzek, John S. "Green Democracy." In *Deliberative Democracy and Beyond: Liberals, Critics, Contestations*, 140–161. Oxford: Oxford University Press, 2000.

Dryzek, John S. *The Politics of the Earth: Environmental Discourses*. Oxford: Oxford University Press, 2005.

Dryzek, John S., Bonnie Honig, and Anne Phillips. "Editors' Introduction." In *The Oxford Handbook of Political Theory*, 3–41. Oxford: Oxford University Press, 2006.

Dryzek, John S., and Hayley Stevenson. "Global Democracy and Earth System Governance." *Ecological Economics* 70, no. 11 (September 2011): 1865–74. doi:10.1016/j.ecolecon.2011.01.021.

Dunlap, Riley E., G. H. Gallup, and A. M. Gallup. *Health of the Planet: Results of a 1992 International Environmental Opinion Survey of Citizens in 24 Nations*. Princeton, NJ: George H. Gallup International Institute, 1993.

Dunlap, Riley E., and Aaron M. McCright. "Organized Climate Change Denial." In *Oxford Handbook of Climate Change and Society*, edited by John S. Dryzek, Richard B. Norgaard, and David Schlosberg, 144–160. Oxford: Oxford University Press, 2011.

Dunlap, Riley E., and Richard York. "The Globalization of Environmental Concern and the Limits of the Postmaterialist Values Explanation: Evidence from Four Multinational Surveys." *Sociological Quarterly* 49, no. 3 (2008): 529–563.

Dunn, James A. "The Politics of Automobility." *The Brookings Review* 17, no. 1 (1999): 40–43.

Dutzik, Tony, and Phineas Baxandall. *A New Direction: Our Changing Relationship with Driving and the Implications for America's Future*. US PIRG Education Fund; Frontier Group, Spring 2013. Accessed March 20, 2014. http://www .uspirg.org/sites/pirg/files/reports/A%20New%20Direction%20vUS.pdf.

Eckersley, Robyn. "Green Politics and the New Class: Selfishness or Virtue?" *Political Studies* 37 (1989): 205–223.

Eckersley, Robyn. *The Green State: Rethinking Democracy and Sovereignty*. Cambridge, MA: MIT Press, 2004.

Epstein, Richard A. *Takings: Private Property and the Power of Eminent Domain*. Cambridge, MA: Harvard University Press, 1985.

Faber, Daniel. *The Struggle for Ecological Democracy: Environmental Justice Movements in the United States*. New York: Guilford Press, 1998.

Fagan, Kevin. "Power on a Shoestring/21-Year Experiment Finds Ways Anyone Can Cut Energy Costs." *San Francisco Chronicle*, January 8, 2001.

First National People of Color Environmental Leadership Summit. "Principles of Environmental Justice," 1991. Accessed December 14. 2012. http://www.ejnet .org/ej/principles.html.

Flacks, Richard. *Making History: The American Left and the American Mind*. New York: Columbia University Press, 1988.

Flint, Anthony. *This Land: The Battle over Sprawl and the Future of America*. Baltimore: Johns Hopkins University Press, 2008.

Ford, Henry, and Samuel Crowther. *My Life and Work*. New York: Doubleday, Page & Company, 1922.

Forgacs, David, ed. *An Antonio Gramsci Reader*. New York: Schocken Books, 1988.

Frank, L. D., M. A. Andresen, and T. L. Schmid. "Obesity Relationships with Community Design, Physical Activity, and Time Spent in Cars." *American Journal of Preventive Medicine* 27, no. 2 (2004): 87–96.

Fraser, Nancy. *Unruly Practices: Power, Discourse and Gender in Contemporary Social Theory*. Minneapolis: University of Minnesota Press, 1989.

Freeden, Michael. *Liberal Languages: Ideological Imaginations and Twentieth-Century Progressive Thought*. Princeton, NJ: Princeton University Press, 2005.

Freudendal-Pedersen, Malene. *Mobility in Daily Life: Between Freedom and Unfreedom*. London: Ashgate, 2009.

Freyfogle, Eric T. *The Land We Share: Private Property and the Common Good*. Washington, DC: Island Press, 2003.

Frumkin, Howard, Lawrence Frank, and Richard J. Jackson. *Urban Sprawl and Public Health: Designing, Planning, and Building for Healthy Communities.* Washington, DC: Island Press, 2004.

Gabrielson, Teena, and Katelyn Parady. "Corporeal Citizenship: Rethinking Green Citizenship through the Body." *Environmental Politics* 19, no. 3 (May 2010): 374–391. doi:10.1080/09644011003690799.

Gabrielson, Teena, Cheryl Hall, John M. Meyer, and David Schlosberg, eds. *Oxford Handbook of Environmental Political Theory.* Oxford: Oxford University Press, forthcoming 2015.

Geuss, Raymond. "Liberalism and Its Discontents." *Political Theory* 30, no. 3 (June 2002): 320–338.

Geuss, Raymond. *Public Goods, Private Goods.* Princeton, NJ: Princeton University Press, 2003.

Gibson, Chris, Gordon Waitt, Lesley Head, and Nick Gill. "Is It Easy Being Green? On the Dilemmas of Material Cultures of Household Sustainability." In *Material Geographies of Household Sustainability*, edited by Ruth Lane and Andrew Gorman-Murray, 19–34. Farnham: Ashgate, 2011.

Giddens, Anthony. *The Constitution of Society: Introduction of the Theory of Structuration.* Oakland: University of California Press, 1984.

Givens, Jennifer E., and Andrew K. Jorgenson. "The Effects of Affluence, Economic Development, and Environmental Degradation on Environmental Concern: A Multilevel Analysis." *Organization & Environment* 24, no. 1 (March 1, 2011): 74–91. doi:10.1177/1086026611406030.

Goodwin, Katherine J. "Reconstructing Automobility: The Making and Breaking of Modern Transportation." *Global Environmental Politics* 10, no. 4 (November 2010): 60–78.

Gorz, André. *Ecology as Politics.* Boston: South End Press, 1980.

Gottlieb, Robert. *Forcing the Spring: The Transformation of the American Environmental Movement.* Washington, DC: Island Press, 1993.

Gottlieb, Robert. *Forcing the Spring: The Transformation of the American Environmental Movement*, revised ed Washington, DC:Island Press, 2005.

Gray, John. "Back to Mill: Review of *The Snake That Swallowed Its Tail: Some Contradictions in Modern Liberalism* by Mark Garnett." *The New Statesman* 17, no. 833 (2004): 50.

Grey, Thomas C. "The Disintegration of Property." In *NOMOS XXII: Property*, edited by J. Roland Pennock and John W. Chapman, 69–85. New York: New York University Press, 1980.

Guber, Deborah L. *The Grassroots of a Green Revolution: Polling America on the Environment.* Cambridge, MA: MIT Press, 2003.

Guha, Ramachandra, and Juan Martinez-Alier. *Varieties of Environmentalism: Essays North and South.* London: Earthscan, 1997.

Hafetz, Jonathan. "'A Man's Home Is His Castle?': Reflections on the Home, the Family, and Privacy During the Late Nineteenth and Early Twentieth Centuries."

William & Mary Journal of Women and the Law 8, no. 2 (February 1, 2002): 175-242.

Hailwood, Simon. *How to Be a Green Liberal: Nature, Value and Liberal Philosophy*. Chesham, UK: Acumen, 2004.

Hall, Cheryl. "What Will It Mean to Be Green? Envisioning Positive Possibilities without Dismissing Loss." *Ethics, Policy & Environment* 16, no. 2 (June 2013): 125–141. doi:10.1080/21550085.2013.801182.

Hanson, Susan. "Gender and Mobility: New Approaches for Informing Sustainability." *Gender, Place & Culture* 17, no. 1 (February 2010): 5–23. doi:10.1080/09663690903498225.

Harden, Blaine. "Anti-Sprawl Laws, Property Rights Collide in Oregon." *Washington Post*, February 28, 2005, 1.

Hardin, Garrett. "The Tragedy of the Commons." *Science* 162 (December 13, 1968): 1243–1248.

Hartz, Louis. *The Liberal Tradition in America: An Interpretation of American Political Thought since the Revolution*. First ed. New York: Harcourt Brace, 1955.

Harvey, David. "The Right to the City." *New Left Review* 53 (October 2008): 23–40.

Hawken, Paul, Amory Lovins, and Hunter Lovins. *Natural Capitalism: Creating the Next Industrial Revolution*. Boston: Little, Brown and Company, 1999

Hawkins, Gay. "Plastic Materialities." In *Political Matter: Technoscience, Democracy, and Public Life*, 119–138. Minneapolis: University of Minnesota Press, 2010.

Hay, Peter. *A Companion to Environmental Thought*. Edinburgh: Edinburgh University Press, 2002.

Hayden, Dolores. *Redesigning the American Dream: Gender, Housing, and Family Life*. Rev. ed. New York: W. W. Norton, 2002.

Hays, Samuel P. *Beauty, Health, and Permanence: Environmental Politics in the United States, 1955–1985*. Cambridge, MA: Cambridge University Press, 1987.

Hays, Samuel P. *Conservation and the Gospel of Efficiency; the Progressive Conservation Movement, 1890–1920*. Cambridge, MA: Harvard University Press, 1959.

Heidegger, Martin. "Building, Dwelling, Thinking." In *Basic Writings*, rev. ed., edited by David Farrell Krell, 343–364. New York: HarperCollins, 1993.

Heilbroner, Robert. *An Inquiry into the Human Prospect: Updated and Reconsidered for the 1980s*. New York: Norton, 1980.

Hickman, Larry A. "Nature as Culture: John Dewey's Pragmatic Naturalism." In *Environmental Pragmatism*, edited by Andrew Light and Erik Katz, 50–72. London: Routledge, 1996.

Hobson, Kersty. "Environmental Politics, Green Governmentality and the Possibility of a 'Creative Grammar' for Domestic Sustainable Consumption." In

Material Geographies of Household Sustainability, edited by Ruth Lane and Andrew Gorman-Murray, 193–210. Farnham: Ashgate, 2011.

Hobson, Kersty. "On the Making of the Environmental Citizen." *Environmental Politics* 22, no. 1 (February 2013): 56–72. doi:10.1080/09644016.2013.755388.

Hohfeld, Wesley Newcomb. "Fundamental Legal Conceptions as Applied in Judicial Reasoning." *Yale Law Journal* 26, no. 8 (1917): 710–770.

Hohfeld, Wesley Newcomb. "Some Fundamental Legal Conceptions as Applied in Judicial Reasoning." *Yale Law Journal* 23, no. 1 (1913): 16–59.

Holland, Breena. "Ecological Constraints and Value-Pluralism: Why Democracies Should Prohibit Some Ways of Life." Paper presented at the Western Political Science Association annual meeting, Oakland, CA, 2005.

Honoré, A.M. "Ownership." In *Oxford Essays in Jurisprudence*, edited by A. G. Guest, 107–147. Oxford: Oxford University Press, 1961.

Horwitz, Morton J. *The Transformation of American Law 1780–1860*. Cambridge, MA: Harvard University Press, 1977.

Hulme, Mike. *Why We Disagree about Climate Change: Understanding Controversy, Inaction, and Opportunity*. Cambridge: Cambridge University Press, 2009.

Hunter, David B. "An Ecological Perspective on Property: A Call for Judicial Protection of the Public's Interest in Environmentally Critical Resources." *Harvard Environmental Law Review* 12 (1988): 311–383.

Hyde, Lewis. *The Gift: Imagination and the Erotic Life of Property*. New York: Random House, 1983.

Inglehart, Ronald. "Changing Values among Western Publics from 1970–2006." *West European Politics* 31, nos. 1–2 (2008): 130–146.

Inglehart, Ronald. *Culture Shift in Advanced Industrial Society*. Princeton, NJ: Princeton University Press, 1990.

Inglehart, Ronald. "Public Support for Environmental Protection: Objective Problems and Subjective Values in 43 Societies." *PS: Political Science and Politics* 28, no. 1 (March 1995): 57–72.

Inglehart, Ronald. *The Silent Revolution: Changing Values and Political Styles Among Western Publics*. Princeton, NJ: Princeton University Press, 1977.

Ingraham, Christopher. "The Green Movement Has a Millennial Problem." *Washington Post*, March 7, 2014. Accessed May 12, 2014. http://www
.washingtonpost.com/blogs/wonkblog/wp/2014/03/07/the-green-movement
-has-a-millennial-problem/.

"The Ivan Illich Archive—Energy and Equity," n.d. Accessed July 29, 2011. http://
www.preservenet.com/theory/Illich/EnergyEquity/Energy%20and%20Equity.
htm.

Jacobs, Harvey M. "The Future of an American Ideal." In *Private Property in the 21st Century*, edited by Harvey M. Jacobs, 171–184. Northampton, MA: Edward Elgar, 2004.

Jacobsen, Linda A., Mark Mather, and Genevieve Dupuis. "Household Change in the United States." *Population Reference Bureau*, September 2012. Accessed March 13, 2013. http://www.prb.org/Publications/PopulationBulletins/2012/us-household-change.aspx.

Jacoby, Karl. *Crimes against Nature: Squatters, Poachers, Thieves, and the Hidden History of American Conservation*. Berkeley: University of California Press, 2001.

Jensen, Derrick. *Endgame, Volume 1: The Problem of Civilization*. New York: Seven Stories Press, 2005.

Johnson, Elmer W. "Taming the Car and Its User: Should We Do Both?" *Bulletin of the American Academy of Arts and Sciences* 46, no. 2 (November 1, 1992): 13–29.

Johnsson-Latham, Gerd. "A Study on Gender Equality as a Prerequisite for Sustainable Development." *Report to the Environment Advisory Council*. 2007. Accessed March 19, 2014. http://www.uft.uni-bremen.de/oekologie/hartmutkoehler_fuer_studierende/MEC/09-MEC-reading/gender%202007%20EAC%20rapport_engelska.pdf.

Kaika, Maria. "Interrogating the Geographies of the Familiar: Domesticating Nature and Constructing the Autonomy of the Modern Home." *International Journal of Urban and Regional Research* 28, no. 2 (2004): 265–286. doi: 10.1111/j.0309-1317.2004.00519.x.

Karlsson, Rasmus. "Ambivalence, Irony, and Democracy in the Anthropocene." *Futures* 46 (February 2013): 1–9. doi:10.1016/j.futures.2012.12.002.

Karlsson, Rasmus. "Individual Guilt or Collective Progressive Action? Challenging the Strategic Potential of Environmental Citizenship Theory." *Environmental Values* 21, no. 4 (November 1, 2012): 459–474. doi:10.3197/0963271 12X13466893628102.

Kassiola, Joel J. *The Death of Industrial Civilization: The Limits to Economic Growth and the Repoliticization of Advanced Industrial Society*. Albany: SUNY Press, 1990.

Katz, Ashley. "About LEED." *US Green Building Council*, July 25, 2012. Accessed July 23, 2013. http://www.usgbc.org/articles/about-leed.

Kaufmann, V., M. M. Bergman, and D. Joye. "Motility: Mobility as Capital." *International Journal of Urban and Regional Research* 28, no. 4 (2004): 745–756.

Keat, Russell. "Citizens, Consumers and the Environment: Reflections on *The Economy of the Earth*." *Environmental Values* 3, no. 4 (November 1, 1994): 333–349. doi:10.3197/096327194776679674.

Kempton, Willett M., James S. Boster, and Jennifer A. Hartley. *Environmental Values in American Culture*. Cambridge, MA: MIT Press, 1995.

Kim, So Young, and Yael Wolinsky-Nahmias. "Cross-National Public Opinion on Climate Change: The Effects of Affluence and Vulnerability." *Global Envi-*

ronmental Politics 14, no. 1 (February 2014): 79–106. doi:10.1162/GLEP_a _00215.

Kogl, Alexandra. "A Hundred Ways of Beginning: The Politics of Everyday Life." *Polity* 41, no. 4 (2009): 514–535.

Kotkin, Joel. "The Triumph of Suburbia." *Newgeography.com*, April 29, 2013. Accessed July 15, 2013. http://www.newgeography.com/content/003667-the -triumph-suburbia.

Kovel, Joel. *The Enemy of Nature: The End of Capitalism or the End of the World?* New York: Zed, 2002.

Krause, Sharon R. "Bodies in Action: Corporeal Agency and Democratic Politics." *Political Theory* 39, no. 3 (March 2011): 299–324. doi:10.1177/ 0090591711400025.

Kunstler, J. H. *The Geography of Nowhere: The Rise and Decline of America's Man-Made Landscape.* New York: Simon and Schuster, 1993.

Kuttner, Robert. "Share Economy or Bare Economy?" *Huffington Post*, April 27, 2014. Accessed April 28, 2014. http://www.huffingtonpost.com/robert-kuttner/ share-economy-or-bare-eco_b_5223968.html.

Latour, Bruno. "Why Has Critique Run out of Steam? From Matters of Fact to Matters of Concern." *Critical Inquiry* 30, no. 2 (2004): 225–248.

Latour, Bruno. *Politics of Nature: How to Bring the Sciences into Democracy.* Cambridge, MA: Harvard University Press, 2004.

Latour, Bruno. *Reassembling the Social: An Introduction to Actor-Network-Theory.* Cambridge: Cambridge University Press, 2007.

Latour, Bruno. *We Have Never Been Modern.* Cambridge, MA: Harvard University Press, 1993.

LaVaque-Manty, Mika. *Arguments and Fists: Political Agency and Justification in Liberal Theory.* New York: Routledge, 2002.

Lear, Jonathan. *Radical Hope: Ethics in the Face of Cultural Devastation.* Cambridge, MA: Harvard University Press, 2006.

Lemann, Nicholas. "When the Earth Moved." *The New Yorker*, April 15, 2013. Accessed May 16, 2013. http://www.newyorker.com/arts/critics/atlarge/2013/ 04/15/130415crat_atlarge_lemann.

Leopold, Aldo. "[1947] Forward." In *Companion to a Sand County Almanac*, edited by J. Baird Callicott, 281–290. Madison: University of Wisconsin Press, 1987.

Leopold, Aldo. *A Sand County Almanac: And Sketches Here and There.* Oxford: Oxford University Press, 1949.

Lilley, Sasha, David McNally, Eddie Yuen, and James Davis. *Catastrophism: The Apocalyptic Politics of Collapse and Rebirth.* Oakland, CA: PM Press, 2012.

Locke, John. *Two Treatises of Government.* Edited by Peter Laslett. Cambridge: Cambridge University Press, 1988.

Lomasky, Loren E. "Autonomy and Automobility." *Independent Review* 2, no. 1 (1997): 5–28.

Lovelock, James E. *The Vanishing Face of Gaia: The Final Warning*. New York: Basic Books, 2009.

Luke, Timothy W. "Deep Ecology as Political Philosophy." In *Eco-Critique: Contesting the Politics of Nature, Economy, and Culture*, 1–27. Minneapolis: University of Minnesota Press, 1997.

Luke, Timothy W. *Ecocritique: Contesting the Politics of Nature, Economy, and Culture*. Minneapolis: University of Minnesota Press, 1997.

Luke, Timothy W. "The People, Politics, and the Planet: Who Knows, Protects, and Serves Nature Best?" In *Democracy and the Claims of Nature*, edited by Ben A. Minteer and Bob Pepperman Taylor, 301–320. Lanham, MD: Rowman and Littlefield, 2002.

MacGregor, Sherilyn. *Beyond Mothering Earth: Ecological Citizenship and the Politics of Care*. Vancouver: UBC Press, 2011.

MacGregor, Sherilyn. "Ecological Citizenship." In *Handbook of Political Citizenship and Social Movements*, edited by H.-A. van der Heijden, 107–132. Cheltenham, UK: Edward Elgar, 2014.

Machan, Tibor R. "Pollution and Political Theory." In *Earthbound: New Introductory Essays in Environmental Ethics*, edited by Tom Regan, 74–106. New York: Random House, 1984.

MacPherson, C. B. "The Meaning of Property." In *Property: Mainstream and Critical Positions*, 1–13. Toronto: University of Toronto Press, 1978.

Maniates, Michael. "Individualization Plant a Tree, Buy a Bike, Save the World?" In *Confronting Consumption*, edited by Thomas Princen, Michael Maniates, and Ken Conca, 43–66. Cambridge, MA: MIT Press, 2002.

Maniates, Michael, and John M. Meyer, eds. *The Environmental Politics of Sacrifice*. Cambridge, MA: MIT Press, 2010.

Manno, Jack. "Commoditization: Consumption Efficiency and an Economy of Care and Connection." In *Confronting Consumption*, edited by Thomas Princen, Michael Maniates, and Ken Conca, 67–100. Cambridge, MA: MIT Press, 2002.

Marres, Noortje. "Front-Staging Nonhumans: Publicity as a Constraint on the Political Activity of Things." In *Political Matter: Technoscience, Democracy, and Public Life*, edited by Bruce Braun and Sarah Whatmore, 177–209. Minneapolis: University of Minnesota Press, 2010.

Marres, Noortje. "Issues Spark a Public into Being: A Key but Often Forgotten Point of the Lippmann-Dewey Debate." In *Making Things Public*, edited by Bruno Latour and Peter Weibel, 208–217. Cambridge, MA: MIT Press, 2005.

Marres, Noortje. *Material Participation: Technology, the Environment and Everyday Publics*. London: Palgrave Macmillan, 2012.

Marres, Noortje. "Testing Powers of Engagement: Green Living Experiments, the Ontological Turn and the Undoability of Involvement." *European Journal of Social Theory* 12, no. 1 (February 1, 2009): 117–133. doi:10.1177/ 1368431008099647.

Martin, Sheila A., Meg Merrick, Erik Rundell, and Katie Shriver. "What Is Driving Measure 37 Claims in Oregon?" Paper presented at the Urban Affairs Association annual meeting, Seattle, WA, April 26, 2007. Accessed March 19, 2014. http://www.pdx.edu/sites/www.pdx.edu.ims/files/ims_M37April07UAAppt .pdf.

Marx, Karl. "On the Jewish Question." In *The Marx-Engels Reader*, edited by Robert C. Tucker, 26–52. New York: W. W. Norton, 1978.

Marx, Karl, and Friedrich Engels. *The Marx-Engels Reader*. Edited by Robert C. Tucker. New York: W. W. Norton, 1978.

Marx, Leo. "Technology: The Emergence of a Hazardous Concept." *Social Research* 64, no. 3 (Fall 1997): 965–989.

Maslow, Abraham H. "A Theory of Human Motivation." *Psychological Review* 50, no. 4 (July 1943): 370–396. doi:10.1037/h0054346.

Mather, Mark, Kelvin Pollard, and Linda A. Jacobsen. *Reports America*. Washington, DC: Population Reference Bureau, 2009.

May, Timothy D. "Shared Housing: The Sharing Economy Gives Roommates a New Image." *Christian Science Monitor*, March 30, 2014. Accessed August 19, 2014. http://www.csmonitor.com/The-Culture/Family/2014/0330/Shared -housing-The-sharing-economy-gives-roommates-a-new-image.

McDonough, William, and Michael Braungart. *Cradle to Cradle: Remaking the Way We Make Things*. New York: North Point Press, 2002.

Meaton, Julia, and David Morrice. "The Ethics and Politics of Private Automobile Use." *Environmental Ethics* 18, no. 1 (1996): 39–54.

Mellor, Mary. *Feminism and Ecology*. Cambridge, UK: Polity Press, 1997.

Meltzer, Graham. *Sustainable Community: Learning from the Cohousing Model*. Bloomington, IN: Trafford Publishing, 2005.

Mertig, Angela G., and Riley E. Dunlap. "Environmentalism, New Social Movements, and the New Class: A Cross-National Investigation." *Rural Sociology* 66, no. 1 (2001): 113–136.

Meyer, John M. "A Democratic Politics of Sacrifice?" In *The Environmental Politics of Sacrifice*, edited by Michael Maniates and John M. Meyer, 13–32. Cambridge, MA: MIT Press, 2010.

Meyer, John M. "Global Liberalism, Environmentalism, and the Changing Boundaries of the Political: Karl Polanyi's Insights." In *Environmental Values in a Globalizing World: Nature, Justice, and Governance*, edited by Jouni Paavola and Ian Lowe, 83–101. London: Routledge, 2005.

Meyer, John M. "Green Liberalism and Beyond." *Organization and Environment* 18, no. 1 (March 2005): 116–120.

Meyer, John M. *Political Nature: Environmentalism and the Interpretation of Western Thought*. Cambridge, MA: MIT Press, 2001.

Meyer, John M. "Political Theory and the Environment." In *Oxford Handbook of Political Theory*, edited by John S. Dryzek, Bonnie Honig, and Anne Phillips, 773–791. Oxford: Oxford University Press, 2006.

Meyer, John M. "Populism, Paternalism, and the State of Environmentalism in the U.S." *Environmental Politics* 17, no. 2 (April 2008): 219–236.

Meyer, John M. "Review of *Requiem for Modern Politics: The Tragedy of the Enlightenment and the Challenge of the New Millennium*, by William Ophuls." *Political Science* 51, no. 1 (July 1999) 87–88.

Meyer, John M. "Review of *The Promise of Green Politics: Environmentalism and the Public Sphere* by Douglas Torgerson." *American Political Science Review* 94, no. 1 (March 2000): 181–182.

Millard-Ball, Adam, and Lee Schipper. "Are We Reaching Peak Travel? Trends in Passenger Transport in Eight Industrialized Countries." *Transport Reviews* 31, no. 3 (May 2011): 357–378. doi:10.1080/01441647.2010.518291.

Mock, Brentin. "Mainstream Green Is Still Too White." *COLORLINES*, April 2, 2013. Accessed April 17, 2013. http://colorlines.com/archives/2013/04/message_from_the_grassroots_dont_blow_it_on_climate_change_this_time.html.

Mortenson, Eric. "Started with Measure 37, Oregon Land-Use War Settled with a Muted Impact on the Land." *The Oregonian*, February 1, 2011. http://www.oregonlive.com/environment/index.ssf/2011/02/oregon_land-use_war_gets_settl.html.

Mortenson, Eric. "Voters Approve Land-Use Rules Changes." *The Oregonian*, November 6, 2007.

Moynihan, Michael C. "Driven Crazy." *Reason*, November 2009. http://reason.com/archives/2009/11/03/driven-crazy.

Mumford, Lewis. *The City in History: Its Origins, Its Transformations, and Its Prospects*. New York: Harcourt Brace Jovanovich, 1961.

Murphy, Andrew R. "Tolerance, Toleration, and the Liberal Tradition." *Polity* 29, no. 4 (1997): 593–623.

Nisbet, Matthew C. "Communicating Climate Change: Why Frames Matter for Public Engagement." *Environment*, April 2009. Accessed January 28, 2012. http://www.environmentmagazine.org/Archives/Back Issues/March-April 2009/Nisbet-full.html.

Nisbet, Matthew C. "Public Opinion and Participation." In *Oxford Handbook of Climate Change and Society*, edited by John S. Dryzek, Richard B. Norgaard, and David Schlosberg, 355–368. Oxford: Oxford University Press, 2011.

Nordhaus, Ted, and Michael Shellenberger. *Break Through: From the Death of Environmentalism to the Politics of Possibility*. Boston: Houghton Mifflin, 2007.

Norton, Brian G. "Ecology and Opportunity: Intergenerational Equity and Sustainable Options." In *Fairness and Futurity: Essays on Environmental Sustainability and Social Justice*, edited by Andrew Dobson, 118–150. Oxford: Oxford University Press, 1999.

Norton, Brian G. *Sustainability: A Philosophy of Adaptive Ecosystem Management*. Chicago: University of Chicago Press, 2005.

Novotny, Patrick. *Where We Live, Work, and Play: The Environmental Justice Movement and the Struggle for a New Environmentalism*. Westport, CT: Praeger, 2000.

Nozick, Robert. *Anarchy, State, and Utopia*. New York: Basic Books, 1974.

Offe, Claus. "Challenging the Boundaries of Institutional Politics: Social Movements since the 1960s." In *Changing Boundaries of the Political*, edited by Charles S. Maier, 63–106. New York: Cambridge University Press, 1987.

Okin, Susan Moller. *Justice, Gender and the Family*. New York: Basic Books, 1989.

Oksanen, Markku. "Environmental Ethics and Concepts of Private Ownership." In *Environmental Ethics and the Global Marketplace*, edited by Dorinda G. Dallmeyer and Albert F. Ike, 114–139. Athens: University of Georgia Press, 1998.

Oksanen, Markku, and Anne-Marie Kumpula. "Is the End of Environmentalism the End of Property? Ownership, the Environment and the Burden of Proof." Paper presented at the ECPR joint sessions: The End of Environmentalism? Turin, 2002.

Ophuls, William. *Ecology and the Politics of Scarcity: Prologue to a Political Theory of the Steady State*. San Francisco: W. H. Freeman, 1977.

Ophuls, William. *Plato's Revenge: Politics in the Age of Ecology*. Cambridge, MA: MIT Press, 2011.

Ophuls, William. *Requiem for Modern Politics: The Tragedy of the Enlightenment and the Challenge of the New Millennium*. Boulder, CO: Westview Press, 1997.

Oppenheimer, Laura. "Only Thing Developed by Measure 37 Is a Headache." *The Oregonian*, March 6, 2005.

Oppenheimer, Laura. "Oregon, Prepare for a Land-Use Fight, Again." *The Oregonian*, March 30, 2007.

Oppenheimer, Laura. "Public Demands Land-Use Clarity." *The Oregonian*, February 23, 2007.

Oppenheimer, Laura. "Voters Nip Libertarian Dreams across U.S." *The Oregonian*, November 13, 2006.

Oregon Department of Land Conservation and Development. *Ballot Measures 37 (2004) and 49 (2007) Outcomes and Effects*, January 2011. Accessed March 19, 2014. http://www.oregon.gov/LCD/docs/publications/m49_2011-01-31.pdf.

Orfield, Myron, and Thomas Luce. "America's Racially Diverse Suburbs: Opportunities and Challenges." *University of Minnesota Law School*. July 20, 2012. http://www.law.umn.edu/uploads/e0/65/e065d82a1c1da0bfef7d86172ec5391e/Diverse_Suburbs_FINAL.pdf.

Palmquist, Matt. "Old without Wheels." *Miller-McCune*, July 14, 2008. Accessed July 28, 2011. http://www.miller-mccune.com/culture-society/old-without-wheels -4419.

Parsons, Adam. "The Sharing Economy: A Short Introduction to Its Political Evolution." *openDemocracy*. March 5, 2014. Accessed 31, 2014. http://www .opendemocracy.net/transformation/adam-parsons/sharing-economy-short -introduction-to-its-political-evolution.

Pateman, Carole. "Feminist Critiques of the Public/Private Dichotomy." In *The Disorder of Women: Democracy, Feminism, and Political Theory*, 118–140. Stanford, CA: Stanford University Press, 1989.

Paterson, Matthew. *Automobile Politics: Ecology and Cultural Political Economy*. Cambridge: Cambridge University Press, 2007.

Paton, G. J. *Seeking Sustainability: On the Prospect of an Ecological Liberalism*. London: Routledge, 2013.

Pearce, Fred. "New Green Vision: Technology as Our Planet's Last Best Hope." *Yale Environment 360*, July 15, 2013. Accessed July 29, 2013. http://e360.yale .edu/feature/new_green_vision_technology_as_our_planets_last_best_hope/ 2671/.

Penn Central Transport Co. v. New York, 438 US 104 (1978).

Pennington, Mark. "Classical Liberalism and Ecological Rationality: The Case for Polycentric Environmental Law." *Environmental Politics* 17, no. 3 (2008): 431–448.

Pepper, David. "Anthropocentrism, Humanism and Eco-Socialism: A Blueprint for the Survival of Ecological Politics." *Environmental Politics* 2, no. 3 (September 1993): 428–452.

Peterson, M. Nils, Tarla Rai Peterson, and Jianguo Liu. *The Housing Bomb: Why Our Addiction to Houses Is Destroying the Environment and Threatening Our Society*. Baltimore: Johns Hopkins University Press, 2013.

Peterson, V. Spike. "Rereading Public and Private: The Dichotomy That Is Not One." *SAIS Review* 20, no. 2 (2000): 11–29.

Pew Research Center for the People and the Press. "GOP Deeply Divided Over Climate Change," November 1, 2013. Accessed May 21, 2014. http://www .people-press.org/files/legacy-pdf/11-1-13%20Global%20Warming%20Release .pdf.

Pitkin, Hanna F. "Justice: On Relating Public and Private." *Political Theory* 9 (1981): 327–352.

Plumwood, Val. "Inequality, Ecojustice, and Ecological Rationality." In *Debating the Earth: The Environmental Politics Reader*, edited by John S. Dryzek and David Schlosberg, 559–583. Oxford: Oxford University Press, 1998.

Polanyi, Karl, "The Economistic Fallacy." In *The Livelihood of Man*, edited by Harry W. Pearson, 5–19. New York: Academic Press, 1977.

Polanyi, Karl. *The Great Transformation: The Political and Economic Origins of Our Time*. Boston: Beacon Press, 1944.

Polanyi, Karl. *The Livelihood of Man*. Edited by Harry W. Pearson. New York: Academic Press, 1977.

Polanyi, Karl. *Primitive, Archaic, and Modern: Economies Essays of Karl Polanyi*. Edited by George Dalton. Boston: Beacon Press, 1968.

Potapov, Alex. "Making Regulatory Takings Reform Work: The Lessons of Oregon's Measure 37." *Environmental Law Reporter News & Analysis* 39 (2009): 10516–10540.

Princen, Thomas. "Review of *Vibrant Matter: A Political Ecology of Things*. By Jane Bennett." *Perspectives on Politics* 9, no. 1 (March 2011): 118–120. doi:10.1017/S1537592710003464.

Prugh, Thomas, Robert Costanza, and Herman E. Daly. *The Local Politics of Global Sustainability*. Washington, DC: Island Press, 2000.

Rajan, Sudhir Chella. "Automobility, Liberalism, and the Ethics of Driving." *Environmental Ethics* 29 (2007): 77–90.

Rajan, Sudhir Chella. *The Enigma of Automobility: Democratic Politics and Pollution Control*. Pittsburgh: University of Pittsburgh Press, 1996.

Reynolds, Julie. "Urban Homesteading: The Integral Urban House." *Mother Earth News*, November/December 1976. Accessed April 2, 2013. http://www.motherearthnews.com/modern-homesteading/urban-homesteading-zmaz76ndztak.aspx.

Richardson, Elmo R. *The Politics of Conservation: Crusades and Controversies, 1897–1913*. Berkeley: University of California Press, 1962.

Rocky Mountain Institute. "Reinventing Fire," n.d. Accessed July 1, 2013. http://www.rmi.org/reinventingfire.

Rose, Carol M. "The Several Futures of Property: Of Cyberspace and Folk Tales, Emission Trades and Ecosystems." *Minnesota Law Review*, no. 83 (1998): 129–182.

Rosenberg, Tina. "It's Not Just Nice to Share, It's the Future." *New York Times*, June 5, 2013. Accessed March 31, 2014. http://opinionator.blogs.nytimes.com/2013/06/05/its-not-just-nice-to-share-its-the-future/.

Rosenthal, Elisabeth. "The End of Car Culture." *New York Times*, sec. Sunday Review, June 29, 2013. Accessed July 1, 2013. http://www.nytimes.com/2013/06/30/sunday-review/the-end-of-car-culture.html.

Rousseau, Jean-Jacques. *Discourse on the Origin of Inequality*. Translated by Donald A. Cress. Indianapolis: Hackett Publishing, 1992.

Ryan, Alan. *Property and Political Theory*. Oxford: Basil Blackwell, 1984.

Rybczynski, Witold. *Home: A Short History of an Idea*. Harmondsworth: Penguin Books, 1987.

Sagoff, Mark. "Can Environmentalists Be Liberals?" In *The Economy of the Earth: Philosophy, Law and the Environment*, 146–170. Cambridge: Cambridge University Press, 1988.

Sagoff, Mark. *The Economy of the Earth: Philosophy, Law and the Environment.* Cambridge: Cambridge University Press, 1988.

Sagoff, Mark. "Free-Market versus Libertarian Environmentalism." *Critical Review* 6, nos. 2–3 (1992): 211–230.

Sagoff, Mark. *Price, Principle, and the Environment.* Cambridge: Cambridge University Press, 2004.

Sagoff, Mark. "Takings, Just Compensation, and the Environment." In *Upstream/ Downstream: Issues in Environmental Ethics*, edited by Donald Scherer, 158–179. Philadelphia: Temple University Press, 1990.

Salleh, Ariel K. "Deeper than Deep Ecology: The Eco-Feminist Connection." *Environmental Ethics* 6 (Winter 1984): 339–345.

Salleh, Ariel K. *Ecofeminism as Politics: Nature, Marx and the Postmodern.* London: Zed Books, 1997.

Sandilands, Catriona. *The Good-Natured Feminist: Ecofeminism and the Quest for Democracy.* Minneapolis: University of Minnesota Press, 1999.

Sandler, Ronald, and Phaedra C. Pezzullo, eds. *Environmental Justice and Environmentalism: The Social Justice Challenge to the Environmental Movement.* Cambridge, MA: MIT Press, 2006.

Santos, Adella, Nancy McGuckin, Kikari Yukiko Nakamoto, Danielle Gray, and Susan Liss. *Summary of Travel Trends 2009 National Household Travel Survey.* US Department of Transportation, June 2011. http://nhts.ornl.gov/2009/pub/stt .pdf.

Saunders, Peter, and Peter Williams. "The Constitution of the Home: Towards a Research Agenda." *Housing Studies* 3, no. 2 (1988): 81–93. doi:10.1080/ 02673038808720618.

Schlosberg, David. *Environmental Justice and the New Pluralism.* New York: Oxford University Press, 1999.

Schlosberg, David. "Theorising Environmental Justice: The Expanding Sphere of a Discourse." *Environmental Politics* 22, no. 1 (2013): 37–55.

Schrank, David, Bill Eisele, and Tim Lomax. *Urban Mobility Report 2012.* Texas Transportation Institute: Texas A&M University System, December 2012. Accessed September 1, 2014. http://mobility.tamu.edu/ums/.

Schuetze, Christopher F. "Environmental Warning Fatigue Sets in." *IHT Rendezvous* (blog), *New York Times*, March 2, 2013. Accessed April 15, 2013. http:// rendezvous.blogs.nytimes.com/2013/03/02/environmental-warning-fatigue -sets-in/.

Schumacher, E. F. *Small Is Beautiful: Economics as if People Mattered.* New York: Harper and Row, 1973.

Seiler, Cotten. *Republic of Drivers: A Cultural History of Automobility in America.* Chicago: University of Chicago Press, 2008.

Sellers, Christopher C. *Crabgrass Crucible: Suburban Nature and the Rise of Environmentalism in Twentieth-Century America*. Chapel Hill: University of North Carolina Press, 2012.

Shear, Michael D. "Obama Tells Donors of Tough Politics of Environment." *New York Times*, sec. U.S./Politics, April 4, 2013. Accessed April 15, 2013. http://www.nytimes.com/2013/04/05/us/politics/obama-donors-keystone-pipeline.html.

Shearman, David J. C., and Joseph Wayne Smith. *The Climate Change Challenge and the Failure of Democracy*. Westport, CT: Praeger Publishers, 2007.

Shellenberger, Michael, and Ted Nordhaus. "The Death of Environmentalism: Global Warming Politics in a Post-Environmental World," 2004. Accessed August 21, 2014. http://thebreakthrough.org/archive/the_death_of_environmentalism.

Sheller, Mimi, and John Urry. "Mobile Transformations of 'Public' and 'Private' Life." *Theory, Culture & Society* 20, no. 3 (June 1, 2003): 107–125. doi:10.1177/02632764030203007.

Sheppard, Steve, ed. *The Selected Writings and Speeches of Sir Edward Coke*. Vol. II. Indianapolis: Liberty Fund, 2003.

Shiffrin, Seana Valentine. "Lockean Arguments for Private Intellectual Property." In *New Essays in the Legal and Political Theory of Property*, edited by Stephen R. Munzer, 138–167. Cambridge: Cambridge University Press, 2001.

Shove, Elizabeth. *Comfort, Cleanliness and Convenience: The Social Organization of Normality*. Oxford, UK: Berg, 2003.

Shove, Elizabeth, Mika Pantzar, and Matt Watson. *The Dynamics of Social Practice: Everyday Life and How It Changes*. London: Sage, 2012.

Singer, Joseph William. *Entitlement: The Paradoxes of Property*. New Haven, CT: Yale University Press, 2000.

Smith, Graham. *Deliberative Democracy and the Environment*. London: Routledge, 2003.

Solnit, Rebecca. "Hope: The Care and Feeding of." *Grist*, August 2, 2011. Accessed January 27, 2014. http://grist.org/living/2011-08-02-hope-the-care-and-feeding-of/.

Spaargaren, Gert. "Theories of Practices: Agency, Technology, and Culture." *Global Environmental Change* 21, no. 3 (August 2011): 813–822. doi:10.1016/j.gloenvcha.2011.03.010.

Sperling, Daniel, and Deborah Gordon. *Two Billion Cars: Driving Toward Sustainability*. New York: Oxford University Press, 2010.

Steger, Manfred B. *Globalism: The New Market Ideology*. Lanham, MD: Rowman and Littlefield, 2002.

Steinberg, Theodore. *Slide Mountain: Or the Folly of Owning Nature*. Berkeley: University of California Press, 1995.

Steinberger, Peter J. "Public and Private." *Political Studies* 47, no. 2 (1999): 292–313. doi:10.1111/1467-9248.00201.

Stephens, Piers H. G. "Green Liberalisms: Nature, Agency and the Good." *Environmental Politics* 10, no. 3 (September 2001): 1–22.

Stern, David I. "The Rise and Fall of the Environmental Kuznets Curve." *World Development* 32, no. 8 (2004): 1419–1439.

Svenvold, Mark. "The New Commute." *Orion Magazine*, October 2014. Accessed September 1, 2014. http://www.orionmagazine.org/index.php/articles/article/8269.

Szasz, Andrew. *Shopping Our Way to Safety: How We Changed from Protecting the Environment to Protecting Ourselves*. Minneapolis: University of Minnesota Press, 2009.

Taylor, Bron R. *Ecological Resistance Movements*. Albany: State University of New York Press, 1995.

Taylor, Dorceta E. "The State of Diversity in Environmental Organizations: Mainstream NGOs, Foundations & Government Agencies." Green 2.0 Working Group, July 2014. Accessed August 21, 2014. http://diversegreen.org/report/.

Taylor, Vanessa, and Frank Trentmann. "Liquid Politics: Water and the Politics of Everyday Life in the Modern City." *Past & Present* 211, no. 1 (May 1, 2011): 199–241. doi:10.1093/pastj/gtq068.

"Thirteen Years of the Public's Top Priorities." Pew Research Center for the People & the Press, January 27, 2014. Accessed June 11, 2013. http://www.people-press.org/interactives/top-priorities/.

Thompson, Allen. "Radical Hope for Living Well in a Warmer World." *Journal of Agricultural and Environmental Ethics* 23, nos. 1–2 (June 17, 2009): 43–59. doi:10.1007/s10806-009-9185-2.

Torgerson, Douglas. "Farewell to the Green Movement? Political Action and the Green Public Sphere." *Environmental Politics* 9, no. 4 (2000): 1–19.

Torgerson, Douglas. *The Promise of Green Politics: Environmentalism and the Public Sphere*. Durham: Duke University Press, 1999.

Trachtenberg, Zev. "Complex Green Citizenship and the Necessity of Judgment." *Environmental Politics* 19, no. 3 (2010): 339–355.

TrueCar. "TrueCar.Com Analyzes Vehicle Registration and Gender Differences." *TrueCar* (blog), June 11, 2010. Accessed March 20, 2014. http://blog.truecar.com/2010/06/11/truecar-com-examines-gender-differences-in-vehicle-registrations/.

Tully, James. *Public Philosophy in a New Key*. Vol. I, *Democracy and Civic Freedom*. Cambridge: Cambridge University Press, 2008.

Underkuffler-Freund, Laura S. "Property: A Special Right." *Notre Dame Law Review*, no. 71 (1996): 1033–1058.

Urry, John. "Automobility, Car Culture and Weightless Travel: A Discussion Paper." Department of Sociology, Lancaster University, January 1999. Accessed August 21, 2014. http://www.lancaster.ac.uk/fass/sociology/research/publications/papers/urry-automobility.pdf.

Urry, John. "The 'System' of Automobility." *Theory, Culture & Society* 21, nos. 4–5 (October 1, 2004): 25–39. doi:10.1177/0263276404046059.

US v. Causby, 328 US 256 (1946).

Valentine, Katie. "The Whitewashing of the Environmental Movement." *Grist*, September 24, 2013. Accessed January 3, 2014. http://grist.org/climate-energy/the-whitewashing-of-the-environmental-movement/.

Vanderheiden, Steve. "Assessing the Case against the SUV." *Environmental Politics* 15, no. 1 (February 2006): 23–40. doi:10.1080/09644010500418688.

Vanderheiden, Steve. *Atmospheric Justice: A Political Theory of Climate Change.* Oxford: Oxford University Press, 2008.

Varner, Gary E. "Environmental Law and the Eclipse of Land as Private Property." In *Ethics and Environmental Policy: Theory Meets Practice*, edited by Frederick Ferre and Peter Hartel, 142–160. Athens: University of Georgia Press, 1994.

Vatn, Arild. "The Environment as a Commodity." *Environmental Values* 9 (2000): 493–509.

Village of Euclid, Ohio v. Ambler Realty Co., 272 US 365 (1926).

Vincent, Andrew. "Liberalism and the Environment." *Environmental Values* 7 (1998): 443–459.

Waldron, Jeremy. *The Right to Private Property.* Oxford: Clarendon Press, 1988.

Walters, Jonathan. "Law of the Land." *Governing Magazine*, May 2005. Accessed August 21, 2014. http://www.governing.com/topics/economic-dev/Law-Land.html.

Walzer, Michael. "The Communitarian Critique of Liberalism." *Political Theory* 18, no. 1 (1990): 6–23.

Walzer, Michael. *Interpretation and Social Criticism.* Cambridge, MA: Harvard University Press, 1993.

Walzer, Michael. "The Political Theory License." *Annual Review of Political Science* 16, no. 1 (2013): 1–9. doi:10.1146/annurev-polisci-032211-214411.

Walzer, Michael. *Spheres of Justice: A Defense of Pluralism and Equality.* New York: Basic Books, 1983.

Warde, Alan. "Consumption and Theories of Practice." *Journal of Consumer Culture* 5, no. 2 (July 1, 2005): 131–153. doi:10.1177/1469540505053090.

Washick, Bonnie, and Elizabeth Wingrove. "Politics That Matter: Thinking about Power and Justice with New Materialists." Paper presented at the Western Political Science Association annual meeting, Portland, OR, 2012.

Weintraub, Jeff. "The Theory and Politics of the Public/Private Distinction." In *Public and Private in Thought and Practice: Perspectives on a Grand Dichotomy*, edited by Jeff Weintraub and Krishan Kumar, 1–42. Chicago: University of Chicago Press, 1997.

Wells, Christopher W. *Car Country: An Environmental History.* Seattle: University of Washington Press, 2012.

Wenar, Leif. "The Concept of Property and the Takings Clause." *Columbia Law Review* 97 (1997): 1923–1946.

Westbrook, Robert B. *John Dewey and American Democracy*. Ithaca, NY: Cornell University Press, 1993.

Weston, Anthony. *Mobilizing the Green Imagination: An Exuberant Manifesto*. Gabriola Island, BC: New Society Publishers, 2012.

Whatmore, Sarah. *Hybrid Geographies: Natures Cultures Spaces*. London: Sage, 2002.

White, Damian Finbar "A Green Industrial Revolution? Sustainable Technological Innovation in a Global Age." *Environmental Politics* 11, no. 2 (June 2002): 1–26. doi:10.1080/714000603.

White, Stephen K. *Sustaining Affirmation: The Strengths of Weak Ontology in Political Theory*. Princeton, NJ: Princeton University Press, 2000.

Whiteside, Kerry H. "The Impasses of Ecological Representation." *Environmental Values* 22, no. 3 (June 1, 2013): 339–358. doi:10.3197/0963271 13X13648087563700.

Williamson, Thad. *Sprawl, Justice, and Citizenship: The Civic Costs of the American Way of Life*. New York: Oxford University Press, 2010.

Willis, Margaret M., and Juliet B. Schor. "Does Changing a Light Bulb Lead to Changing the World? Political Action and the Conscious Consumer." *The ANNALS of the American Academy of Political and Social Science* 644, no. 1 (November 1, 2012): 160–190. doi:10.1177/0002716212454831.

Wilson, James Q. "Cars and Their Enemies." *Commentary*, July 1997, 17–23.

Winner, Langdon. *Autonomous Technology: Technics-out-of-Control as a Theme in Political Thought*. Cambridge, MA: MIT Press, 1977.

Winner, Langdon. *The Whale and the Reactor: A Search for Limits in an Age of High Technology*. Chicago: University of Chicago Press, 1986.

Wissenburg, Marcel. *Green Liberalism*. London: UCL Press, 1998.

Wissenburg, Marcel. "Liberalism Is Always Greener on the Other Side of Mill: A Reply to Piers Stephens." *Environmental Politics* 10, no. 3 (2001): 23–42.

Wissenburg, Marcel. "Political Appeasement and Academic Critique: The Case of Environmentalism." *Philosophy & Social Criticism*, 39, no. 7 (2013): 675–691. doi:10.1177/0191453713491233.

Wissenburg, Marcel. "Sustainability and the Limits of Liberalism." In *Sustaining Liberal Democracy*, edited by John Barry and Marcel Wissenburg, 192–204. New York: Palgrave, 2001.

Worster, Donald. *Nature's Economy: A History of Ecological Ideas*. Cambridge: Cambridge University Press, 1985.

Yack, Bernard. *The Fetishism of Modernities: Epochal Self-Consciousness in Contemporary Social and Political Thought*. Notre Dame, IN: University of Notre Dame Press, 1997.

Yack, Bernard. "Liberalism and Its Communitarian Critics: Does Liberal Practice 'Live Down' to Liberal Theory?" In *Community in America: The Challenges of Habits of the Heart*, edited by Charles H. Reynolds and Ralph V. Norman, 149–167. Berkeley: University of California Press, 1988.

Yack, Bernard. *The Longing for Total Revolution: Philosophic Sources of Social Discontent from Rousseau to Marx and Nietzsche*. Princeton, NJ: Princeton University Press, 1986.

Yack, Bernard. *The Problems of a Political Animal: Community, Justice, and Conflict in Aristotelian Political Thought*. Berkeley: University of California Press, 1993.

Young, Iris Marion. "House and Home: Feminist Variations on a Theme." In *Feminist Interpretations of Martin Heidegger*, 252–288. University Park: Pennsylvania State University Press, 2001.

Index